W9-AXL-633

STUDENT SUPPLEMENT TO ACCOMPANY ZILL'S

A First Course in Differential Equations with Applications

FOURTH EDITION

Warren S. Wright
LOYOLA MARYMOUNT UNIVERSITY

PWS-KENT Publishing Company
BOSTON

PWS-KENT
Publishing Company

20 Park Plaza
Boston, Massachusetts 02116

PWS-KENT Publishing Company is a division of Wadsworth, Inc.

ISBN 0-534-91569-8

Printed in the United States of America
90 91 92 93 — 10 9 8 7 6 5 4 3 2

Contents

Introduction

The solution of a differential equation frequently involves the integration of some function. For convenience we review here some fundamental techniques of integration. A complete discussion of these and other procedures can be found in any standard calculus text.

SUBSTITUTIONS (Algebraic)

EXAMPLE Evaluate $\int x\sqrt{1+x}\,dx$.

SOLUTION Let $u = 1 + x$ so that $du = dx$. Then

$$\begin{aligned}\int x\sqrt{1+x}\,dx &= \int (u-1)u^{1/2}\,du \\ &= \int (u^{3/2} - u^{1/2})\,du \\ &= \frac{2}{5}u^{5/2} - \frac{2}{3}u^{3/2} + c \\ &= \frac{2}{5}(1+x)^{5/2} - \frac{2}{3}(1+x)^{3/2} + c.\end{aligned}$$

SUBSTITUTIONS (Trigonometric)

EXAMPLE Evaluate $\int \sqrt{1-x^2}\,dx$.

SOLUTION Let $x = \sin\theta$ so that $dx = \cos\theta\,d\theta$. Then

$$\begin{aligned}\int \sqrt{1-x^2}\,dx &= \int (1-\sin^2\theta)^{1/2}\cos\theta\,d\theta \\ &= \int \cos^2\theta\,d\theta.\end{aligned}$$

Recalling the half-angle formula $\cos^2\theta = (1+\cos 2\theta)/2$ we have

$$\begin{aligned}\int \sqrt{1-x^2}\,dx &= \frac{1}{2}\int (1+\cos 2\theta)\,d\theta \\ &= \frac{1}{2}\theta + \frac{1}{4}\sin 2\theta + c \\ &= \frac{1}{2}\theta + \frac{1}{2}\sin\theta\cos\theta + c \\ &= \frac{1}{2}\sin^{-1}x + \frac{1}{2}x\sqrt{1-x^2} + c.\end{aligned}$$

INTEGRATION BY PARTS

We leave it to the reader to verify that the first example above could also be solved using the integration by parts formula

$$\int u\,dv = uv - \int v\,du.$$

EXAMPLE Evaluate $\int x^2 \ln 3x\,dx$.

SOLUTION Let $u = \ln 3x$ and $dv = x^2\,dx$ so that $du = (1/3x)3\,dx = dx/x$ and $v = x^3/3$. It follows that

$$\begin{aligned}\int x^2 \ln 3x\,dx &= \frac{x^3}{3}\ln 3x - \frac{1}{3}\int x^3 \frac{dx}{x} \\ &= \frac{x^3}{3}\ln 3x - \frac{1}{3}\int x^2\,dx \\ &= \frac{x^3}{3}\ln 3x - \frac{1}{9}x^3 + c\end{aligned}$$

PARTIAL FRACTIONS

When integrating a rational function $f(x)/g(x)$, where f and g are polynomials, it is often necessary to decompose the given expression into a sum of rational expressions with linear or quadratic denominators. There are three important cases relating to the factors of $g(x)$:
(a) $g(x)$ can be written entirely in terms of linear factors,
(b) $g(x)$ has linear and irreducible (non-factorable) quadratic factors,
(c) $g(x)$ has repeated linear and quadratic factors.

EXAMPLE Evaluate $\int \frac{x}{x^2+3x+2}\,dx$.

SOLUTION Observe that

$$\int \frac{x}{x^2+3x+2}\,dx = \int \frac{x}{(x+1)(x+2)}\,dx.$$

We now assume a decomposition of the form

$$\frac{x}{x^2+3x+2} = \frac{A}{x+1} + \frac{B}{x+2} \tag{1}$$

where A and B are constants. Putting the right side of (1) over a common denominator and then equating numerators gives

$$x = A(x+2) + B(x+1) \tag{2}$$

or

$$x = (A + B)x + (2A + B).$$

The reader may recall that the last equation is actually an identity; and hence by comparing coefficients of powers of x on both sides of the equality, we obtain the following system of equations for determining A and B.

$$A + B = 1$$

$$2A + B = 0$$

The solution of this system is $A = -1$, $B = 2$. Alternatively, setting $x = -2$ in (2) we obtain $B = 2$; and setting $x = -1$ in (2) we obtain $A = -1$. Note that the values of x are the zeros of the terms multiplying A and B in (2). Now,

$$\int \frac{x}{x^2 + 3x + 2}\,dx = \int \left[-\frac{1}{x+1} + \frac{2}{x+2}\right] dx$$
$$= -\ln|x + 1| + 2\ln|x + 2| + c.$$

EXAMPLE Evaluate $\int \frac{x+2}{x(x^2+1)}\,dx$.

SOLUTION Since $x^2 + 1$ does not factor using real numbers, we assume

$$\frac{x+2}{x(x^2+1)} + \frac{A}{x} = \frac{Bx + C}{x^2 + 1}$$

and therefore

$$x + 2 = A(x^2 + 1) + (Bx + C)x \tag{3}$$
$$= (A + B)x^2 + Cx + A.$$

Setting $x = 0$ in (3) we obtain $A = 2$. Since the denominator has no other real zeros, we now compare the coefficients of x^2 and x:

$$A + B = 0$$

$$C = 1.$$

This system yields $B = -2$ and $C = 1$. Hence,

$$\int \frac{x+2}{x(x^2+1)}\,dx = \int \left[\frac{2}{x} + \frac{-2x + 1}{x^2 + 1}\right] dx$$
$$= 2\int \frac{dx}{x} - \int \frac{2x}{x^2+1}\,dx + \int \frac{dx}{x^2+1}$$
$$= 2\ln|x| - \ln(x^2 + 1) + \tan^{-1} x + c.$$

EXAMPLE Evaluate $\int \frac{2x+1}{(x+2)(x-1)^3}\,dx$.

SOLUTION The factor $x-1$ is repeated three times in the denominator of the integrand. Accordingly, we assume

$$\frac{2x+1}{(x+2)(x-1)^3} = \frac{A}{x+2} + \frac{B}{x-1} + \frac{C}{(x-1)^2} + \frac{D}{(x-1)^3}.$$

Putting the right side over a common denominator and equating numerators implies

$$2x+1 = A(x-1)^3 + B(x+2)(x-1)^2 + C(x+2)(x-1) + D(x+2).$$

Setting $x=-2$ and $x=1$ yields $A=1/9$ and $D=1$, respectively. Now, since the coefficient of x^3 on the right side of the equation is $A+B$ and the coefficient of x^3 on the left side is 0, we have

$$A+B=0.$$

Thus, $B=-1/9$. To solve for C we equate the coefficients of x^2 to obtain

$$-3A+C=0.$$

It follows that $C=1/3$. Finally, we can write

$$\begin{aligned}\int \frac{2x+1}{(x+2)(x-1)^3}\,dx &= \int \left[\frac{1/9}{x+2} - \frac{1/9}{x-1} + \frac{1/3}{(x-1)^2} + \frac{1}{(x-1)^3}\right] dx \\ &= \frac{1}{9}\ln|x+2| - \frac{1}{9}\ln|x-1| - \frac{1}{3}(x-1)^{-1} - \frac{1}{2}(x-1)^{-2} + c.\end{aligned}$$

1 Introduction to Differential Equations

Exercises 1.1

4. The equation can be put into the form

$$x^2\frac{dy}{dx} + (1-x)y = xe^x$$

and hence is linear.

18. First write

$$2xy\,dx + (x^2+2y)\,dy = 0$$

in the form

$$2xy + (x^2+2y)y' = 0.$$

Implicitly differentiating $x^2y + y^2 = c_1$ we obtain

$$x^2y' + 2xy + 2yy' = 0$$

or

$$2xy + (x^2+2y)y' = 0.$$

Since the latter is the given differential equation, we have shown that $x^2y + y^2 = c_1$ is a solution.

22. Writing $y = x|x|$ as

$$y = \begin{cases} x^2, & \text{if } x \geq 0; \\ -x^2, & \text{if } x < 0 \end{cases}$$

we see that

$$|y| = x^2, \quad -\infty < x < \infty,$$

and

$$\sqrt{|y|} = \begin{cases} x, & \text{if } x \geq 0; \\ -x, & \text{if } x < 0. \end{cases}$$

Since

$$y' = \begin{cases} 2x, & \text{if } x \geq 0; \\ -2x, & \text{if } x < 0, \end{cases}$$

it is apparent that $y' = 2\sqrt{|y|}$.

26. Differentiating
$$y = e^{-x^2}\int_0^x e^{t^2}\,dt + c_1 e^{-x^2}$$
we obtain
$$y' = e^{-x^2}e^{x^2} - 2xe^{-x^2}\int_0^x e^{t^2}\,dt - 2c_1 xe^{-x^2}$$
$$= 1 - 2xe^{-x^2}\int_0^x e^{t^2}\,dt - 2c_1 xe^{-x^2}.$$
Substituting into the differential equation, we have
$$y' + 2xy = 1 - 2xe^{-x^2}\int_0^x e^{t^2}\,dt - 2c_1 xe^{-x^2} + 2xe^{-x^2}\int_0^x e^{t^2}\,dt + 2c_1 xe^{-x^2} = 1.$$

36. From $y = x\cos(\ln x)$ we obtain
$$y' = -\sin(\ln x) + \cos(\ln x)$$
and
$$y'' = -\frac{1}{x}\cos(\ln x) - \frac{1}{x}\sin(\ln x).$$
Substituting into the differential equation we have
$$x^2y'' - xy' + 2y = x^2\left[-\frac{1}{x}\cos(\ln x) - \frac{1}{x}\sin(\ln x)\right] - x[-\sin(\ln x) + \cos(\ln x)] + 2x\cos(\ln x)$$
$$= -x\cos(\ln x) - x\sin(\ln x) + x\sin(\ln x) - x\cos(\ln x) + 2x\cos(\ln x)$$
$$= 0.$$

50. The function
$$y = \begin{cases} \sqrt{4 - x^2}, & -2 < x < 0 \\ -\sqrt{4 - x^2}, & 0 \le x < 2 \end{cases}$$
is not continuous at $x = 0$ (the left hand limit is 2 and the right hand limit is -2,) and hence y' does not exist at $x = 0$.

54. Using
$$y' = mx^{m-1} \quad \text{and} \quad y'' = m(m-1)x^{m-2}$$
and substituting into the differential equation we obtain
$$x^2y'' + 6xy' + 4y = [m(m-1) + 6m + 4]x^m.$$
The right side will be zero provided m satisfies
$$m(m-1) + 6m + 4 = m^2 + 5m + 4 = (m+4)(m+1) = 0.$$
Thus, $m = -4, -1$ and two solutions of the differential equation on the interval $0 < x < \infty$ are $y = x^{-4}$ and $y = x^{-1}$.

Exercises 1.2

8. Using implicit differentiation and solving for dy/dx we obtain

$$2c_1x - 2y\frac{dy}{dx} = 0$$

$$\frac{dy}{dx} = \frac{c_1x}{y}. \qquad (1)$$

Solving the original equation for c_1 we find

$$c_1x^2 - y^2 = 1.$$

Substituting into (1) then gives

$$\frac{dy}{dx} = \frac{(1+y^2)/x}{y} = \frac{1+y^2}{xy}$$

16. We compute the first and second derivatives of $y = c_1e^x \cos x + c_2e^x \sin x$:

$$\begin{aligned} y' &= c_1e^x(-\sin x + \cos x) + c_2e^x(\cos x + \sin x) \\ &= -c_1e^x \sin x + c_2e^x \cos x + c_1e^x \cos x + c_2e^x \sin x \\ &= -c_1e^x \sin x + c_2e^x \cos x + y, \\ y'' &= -c_1e^x(\cos x + \sin x) + c_2e^x(-\sin x + \cos x) + y' \\ &= -c_1e^x \cos x - c_2e^x \sin x - c_1e^x \sin x + c_2e^x \cos x + y' \\ &= -y + (y' - y) + y' \\ &= 2y' - 2y. \end{aligned}$$

The differential equation is

$$y'' - 2y' + 2y = 0.$$

28. Differentiating $y^2 = 2x$ gives $2yy' = 2$ or $y' = 1/y$. At a point (x_1, y_1) on the curve, the equation of the tangent line is

$$\begin{aligned} y - y_1 &= \frac{1}{y_1}(x - x_1) \\ &= \frac{1}{y_1}(x - \frac{y_1^2}{2}) \end{aligned}$$

or

$$y = \frac{1}{y_1}x + \frac{y_1}{2}. \tag{2}$$

This is a one-parameter family of straight lines. Differentiating (2) gives

$$\frac{dy}{dx} = \frac{1}{y_1}. \tag{3}$$

Also, equation (2) now yields a quadratic equation for the determination of y_1:

$$(y_1)^2 - 2yy_1 + 2x = 0.$$

It follows that

$$y_1 = y \pm \sqrt{y^2 - 2x}\,.$$

From (3) we find that the differential equation of the family of tangent lines is

$$\frac{dy}{dx} = \frac{1}{y \pm \sqrt{y^2 - 2x}}\,.$$

30. Write the given equation as

$$\left(\frac{y}{x^2}\right)^2 - 2c_1\frac{y}{x^2} - c_2 = 0.$$

By treating this last equation as a quadratic in y/x^2 we find

$$\frac{y}{x^2} = c_1 \pm \sqrt{c_1^2 + c_2}\,.$$

That is,

$$y = k_1x^2 \quad \text{or} \quad y = k_2x^2$$

where $k_1 = c_1 - \sqrt{c_1^2 + c_2}$ and $k_2 = c_1 + \sqrt{c_1^2 + c_2}$. Differentiating and eliminating the parameters k_1 and k_2 gives in either case

$$\frac{dy}{dx} = 2k_1x = 2\left(\frac{y}{x^2}\right)x = \frac{2y}{x}$$

or

$$x\frac{dy}{dx} - 2y = 0.$$

34. We have

decrease in volume of water = volume of water in element of length Δx.

That is,

$$-A_1\Delta h = A_2\Delta x.$$

Dividing by Δt and taking the limit as $\Delta t \to 0$ then gives

$$-A_1 \frac{dh}{dt} = A_2 \frac{dx}{dt}$$

where $dx/dt = v$ is the velocity of the escaping water. Since

potential energy = kinetic energy,

$$mgh = \frac{1}{2} mv^2$$

we have $v = \sqrt{2gh}$. Hence,

$$\frac{dh}{dt} = -\frac{A_2}{A_1}\sqrt{2gh}\,.$$

36. Equating Newton's law with the net forces in the x and y directions gives

$$m\frac{d^2x}{dt^2} = 0$$

and

$$m\frac{d^2y}{dt^2} = -mg$$

respectively.

42. We have from Archimedes' principle

$$\begin{aligned} \text{upward force of water on barrel} &= \text{weight of water displaced} \\ &= (62.4) \times (\text{volume of water displaced}) \\ &= (62.4)\pi(s/2)^2 y \\ &= 15.6\pi s^2 y. \end{aligned}$$

It then follows from Newton's second law that

$$\frac{w}{g}\frac{d^2y}{dt^2} = -15.6\pi s^2 y$$

or

$$\frac{d^2y}{dt^2} + \frac{15.6\pi s^2 g}{w} y = 0,$$

where $g = 32$ and w is the weight of the barrel in pounds.

2 First-Order Differential Equations

Exercises 2.1

8. Write the differential equation in the form

$$y' = \frac{y+x}{y-x}.$$

Identifying $f(x,y) = \dfrac{y+x}{y-x}$ we have

$$\frac{\partial f}{\partial y} = -\frac{2x}{(y-x)^2}.$$

By Theorem 2.1 then, the differential equation will have a unique solution in any rectangular region of the plane where $y \neq x$; that is, in any rectangular region that does not intersect the line $y = x$.

14. A function satisfying the differential equation and the initial condition is $y = 1$. Although $f(x,) = |y-1|$ is continuous, $\partial f/\partial y$ is not continuous at $y = 1$, so Theorem 2.1 does not apply.

16. (a) Since $1 + y^2$ and its partial derivative with respect to y are continuous everywhere in the plane, the differential equation has a unique solution through every point in the plane.

    (b) Since $\frac{d}{dx}(\tan x) = \sec^2 x = 1 + \tan^2 x$ and $\tan 0 = 0$, $y = \tan x$ satisfies the differential equation and the initial condition.

    (c) Since $-2 < \pi/2 < 2$ and $\tan x$ is undefined for $x = \pi/2$, $y = \tan x$ is not a solution on the interval $-2 < x < 2$.

    (d) Since $\tan x$ is differentiable and continuous on $-1 < x < 1$, $y = \tan x$ is a solution of the initial value problem on the interval $-1 < x < 1$.

Exercises 2.2

14. We first multiply both sides of the equation by dx:

$$e^x y\,dy = (e^{-y} + e^{-2x}e^{-y})dx.$$

Factoring and separating variables, we have

$$e^x y\,dy = e^{-y}(1 + e^{-2x})dx$$
$$ye^y\,dy = \frac{1 + e^{-2x}}{e^x}\,dx.$$

Integrating,

$$\int ye^y\,dy = \int (e^{-x} + e^{-3x})dx$$
$$ye^y - e^y = -e^{-x} - \frac{1}{3}e^{-3x} + c.$$

24. We first subtract N from both sides and factor.

$$\frac{dN}{dt} = N(te^{t+2} - 1)$$
$$\int \frac{dN}{N} = \int (te^{t+2} - 1)dt$$
$$\ln |N| = te^{t+2} - e^{t+2} - t + c$$

32. We add $1/y$ to both sides.

$$2\frac{dy}{dx} = \frac{2x+1}{y}.$$

Separating variables and integrating, we have

$$\int 2y\,dy = \int (2x + 1)dx$$
$$y^2 = x^2 + x + c.$$

36. Using the formulas for the sine of the sum and difference of two angles, we have

$$\sec y\,\frac{dy}{dx} + \sin x \cos y - \cos x \sin y = \sin x \cos y + \cos x \sin y$$

$$\sec y \frac{dy}{dx} = 2\cos x \sin y$$
$$\frac{dy}{2\sin y \cos y} = \cos x\, dx$$
$$\frac{dy}{\sin 2y} = \cos x\, dx$$
$$\int \csc 2y\, dy = \int \cos x\, dx$$
$$\frac{1}{2}\ln|\csc 2y - \cot 2y| = \sin x + c.$$

40. Separating variables we have

$$\int \frac{dy}{y+\sqrt{y}} = \int \frac{dx}{x+\sqrt{x}}.$$

To integrate $dx/(x+\sqrt{x})$ make the substitution $u^2 = x$. Then $2u\,du = dx$ and

$$\int \frac{dx}{x+\sqrt{x}} = \int \frac{2u\,du}{u^2+u} = \int \frac{2\,du}{u+1} = 2\ln|u+1| + c = 2\ln(\sqrt{x}+1) + c.$$

Thus

$$\int \frac{dy}{y+\sqrt{y}} = \int \frac{dx}{x+\sqrt{x}},$$

which gives

$$2\ln(\sqrt{y}+1) = 2\ln(\sqrt{x}+1) + c$$
$$\ln(\sqrt{y}+1) = \ln(\sqrt{x}+1) + c_1$$
$$e^{\ln(\sqrt{y}+1)} = e^{\ln(\sqrt{x}+1)+c_1}$$
$$\sqrt{y}+1 = e^{c_1}(\sqrt{x}+1)$$
$$\sqrt{y} = c_2(\sqrt{x}+1) - 1$$
$$y = [c_2(\sqrt{x}+1) - 1]^2.$$

48. Writing $y' = dy/dx$ and separating variables we have

$$\frac{dy}{dx} = 1 - 2y$$
$$\int \frac{dy}{1-2y} = \int dx$$
$$-\frac{1}{2}\ln|1-2y| = x + c$$
$$\ln|1-2y| = -2x + c_1$$
$$|1-2y| = e^{-2x+c_1}.$$

Letting the constant c_2 absorb the absolute value symbol,

$$\begin{aligned} 1 - 2y &= c_2 e^{-2x} \\ -2y &= c_2 e^{-2x} - 1 \\ y &= \frac{1}{2} - c_3 e^{-2x}. \end{aligned}$$

Using the initial condition, we set $x = 0$ and $y = 5/2$. Then

$$\begin{aligned} \frac{5}{2} &= \frac{1}{2} - c_3 e^0 \\ 2 &= -c_3. \end{aligned}$$

Thus

$$y = \frac{1}{2} + 2e^{-2x}.$$

50. Multiplying by dx and separating variables, we have

$$\begin{aligned} x\,dy &= (y^2 - y)dx \\ \frac{dy}{y^2 - y} &= \frac{dx}{x} \qquad (1) \\ \frac{dy}{y(y-1)} &= \frac{dx}{x}. \end{aligned}$$

Using partial fractions,

$$\begin{aligned} \int\left(-\frac{1}{y} + \frac{1}{y-1}\right) dy &= \int \frac{dx}{x} \\ -\ln|y| + \ln|y-1| &= \ln|x| + c \\ \ln\left|\frac{y-1}{y}\right| &= \ln|x| + c \\ \left|\frac{y-1}{y}\right| &= e^{\ln|x|+c} \\ \frac{y-1}{y} &= c_1 x \\ y - 1 &= c_1 xy \\ y - c_1 xy &= 1 \\ y(1 - c_1 x) &= 1 \\ y &= \frac{1}{1 - c_1 x}. \end{aligned}$$

(a) Setting $x = 0$ and $y = 1$ we obtain

$$1 = \frac{1}{1-0} = 1.$$

Any value of c_1 will work, so let $c_1 = 0$. Then $y = 1$ is a solution through $(0, 1)$.

(b) Setting $x = y = 0$ we obtain

$$0 = \frac{1}{1-0} = 1,$$

which shows that there is no solution of the form $y = 1/(1 - c_1 x)$ through $(0, 0)$. Looking back through the steps of the solution of the differential equation, we see that in (1) y became a factor in the denominator. Therefore, $y = 0$ may be considered as a special case. This is, in fact, a solution through $(0, 0)$.

(c) Setting $x = y = 1/2$ we obtain

$$\frac{1}{2} = \frac{1}{1 - c_1/2} = \frac{2}{2 - c_1}.$$

Thus, $c_1 = -2$ and $y = 1/(1 + 2x)$ is a solution through $(1/2, 1/2)$.

Exercises 2.3

8. Since

$$\begin{aligned} f(tx, ty) &= \frac{\ln(tx)^3}{\ln(ty)^3} \\ &= \frac{3\ln tx}{3\ln ty} \\ &\neq t^n \frac{\ln x}{\ln y} \end{aligned}$$

we see that the function is not homogeneous.

20. Write the equation as

$$x\,dy = (y + \sqrt{x^2 + y^2}\,)dx.$$

Now let $y = ux$:

$$\begin{aligned} x(u\,dx + x\,du) &= (ux + \sqrt{x^2 + u^2x^2}\,)dx \\ x^2 du &= x\sqrt{1 + u^2}\,dx \\ \frac{du}{\sqrt{1 + u^2}} &= \frac{dx}{x} \\ \int \frac{du}{\sqrt{1 + u^2}} &= \int \frac{dx}{x} \\ \sinh^{-1} u &= \ln|x| + c. \end{aligned}$$

Recall from calculus that $\sinh^{-1} u = \ln(u + \sqrt{u^2+1})$. Thus, the preceding equation can be written as

$$\begin{aligned} \ln(u + \sqrt{u^2+1}) &= \ln|x| + \ln c_1 \\ &= \ln c_1 x \end{aligned}$$

and so

$$\begin{aligned} u + \sqrt{u^2+1} &= c_1 x \\ \frac{y}{x} + \sqrt{\frac{y^2}{x^2}+1} &= c_1 x \\ y + \sqrt{y^2+x^2} &= c_1 x^2. \end{aligned}$$

26. Letting $y = ux$ we have

$$\begin{aligned} (x^2 e^{-u} + u^2 x^2)\, dx &= ux^2(u\, dx + x\, du) \\ x^2 e^{-u} dx &= ux^3 du \\ \int \frac{dx}{x} &= \int ue^u du \\ \ln|x| &= ue^u - e^u + c \\ \ln|x| &= \frac{y}{x} e^{y/x} - e^{y/x} + c \\ x \ln|x| &= e^{y/x}(y - x) + cx. \end{aligned}$$

30. Letting $y = ux$ we have

$$\begin{aligned} (x^2 + ux^2 + 3u^2x^2)\, dx &= (x^2 + 2ux^2)(u\, dx + x\, du) \\ x^2(1 + u + 3u^2)\, dx &= x^2(1 + 2u)(u\, dx + x\, du) \\ (1 + u + 3u^2)\, dx &= (u + 2u^2)dx + x(1 + 2u)du \\ (1 + u^2)\, dx &= x(1 + 2u)du \\ \frac{dx}{x} &= \frac{1 + 2u}{1 + u^2} du \\ \int \frac{dx}{x} &= \int \frac{du}{1+u^2} + \int \frac{2u}{1+u^2} du \\ \ln|x| &= \tan^{-1} u + \ln(1 + u^2) + c \\ \ln|x| &= \tan^{-1} \frac{y}{x} + \ln\left(1 + \frac{y^2}{x^2}\right) + c \end{aligned}$$

$$\ln|x| = \tan^{-1}\frac{y}{x} + \ln(x^2+y^2) - \ln x^2 + c$$

$$3\ln|x| = \tan^{-1}\frac{y}{x} + \ln(x^2+y^2) + c.$$

36. Letting $x = vy$ we have

$$y(v\,dy + y\,dv) + (y\cos v - vy)\,dy = 0$$

$$y^2 dv + y\cos v\,dy = 0$$

$$\frac{dv}{\cos v} + \frac{dy}{y} = 0$$

$$\sec v\,dv + \frac{dy}{y} = 0$$

$$\int \sec v\,dv + \int \frac{dy}{y} = 0$$

$$\ln|\sec v + \tan v| + \ln|y| = \ln c$$

$$y(\sec\frac{x}{y} + \tan\frac{x}{y}) = c_1.$$

Substituting $x = 0$ and $y = 2$ in the last equation yields $c_1 = 2$. Hence,

$$y\left(\sec\frac{x}{y} + \tan\frac{x}{y}\right) = 2.$$

44. Letting $y = ux$ we have

$$(u\,dx + x\,du) - u\,dx = \cosh u\,dx$$

$$x\,du = \cosh u\,dx$$

$$\frac{du}{\cosh u} = \frac{dx}{x}$$

$$\int \operatorname{sech} u\,du = \int \frac{dx}{x}$$

$$\tan^{-1}(\sinh u) = \ln|x| + c$$

$$\tan^{-1}\left(\sinh\frac{y}{x}\right) = \ln|x| + c.$$

Substituting $x = 1$ and $y = 0$ gives $c = 0$. Thus, the solution of the initial value problem is

$$\tan^{-1}\left(\sinh\frac{y}{x}\right) = \ln|x|.$$

Exercises 2.4

8. Write the equation in the form

$$\left(1 + \ln x + \frac{y}{x}\right) dx + (-1 + \ln x)\, dy = 0$$

and identify

$$M = 1 + \ln x + \frac{y}{x} \qquad \text{and} \qquad N = -1 + \ln x.$$

Since

$$\frac{\partial M}{\partial y} = \frac{1}{x} = \frac{\partial N}{\partial x}$$

the equation is exact. Therefore

$$\frac{\partial f}{\partial y} = -1 + \ln x$$

and

$$f = -y + y \ln x + h(x).$$

Now

$$\frac{\partial f}{\partial x} = \frac{y}{x} + h'(x) = M = 1 + \ln x + \frac{y}{x}$$

and so

$$h'(x) = 1 + \ln x.$$

Using integration by parts, we find

$$h(x) = x \ln x.$$

Hence, the solution of the differential equation is

$$-y + y \ln x + x \ln x = c.$$

16. Write the equation in the form

$$(xy^2 \sinh x + y^2 \cosh x)\, dx + (e^y + 2xy \cosh x)\, dy = 0$$

and identify

$$M = xy^2 \sinh x + y^2 \cosh x \qquad \text{and} \qquad N = e^y + 2xy \cosh x.$$

Since

$$\frac{\partial M}{\partial y} = 2xy \sinh x + 2y \cosh x = \frac{\partial N}{\partial x}$$

the equation is exact. Therefore,

$$\frac{\partial f}{\partial y} = e^y + 2xy \cosh x$$

and

$$f = e^y + xy^2 \cosh x + h(x).$$

Now

$$\frac{\partial f}{\partial x} = xy^2 \sinh x + y^2 \cosh x + h'(x) = M = xy^2 \sinh x + y^2 \cosh x$$

which implies $h'(x) = 0$ or $h = 0$. Hence, the solution of the differential equation is

$$e^y + xy^2 \cosh x = c.$$

22. Write the equation in the form

$$(2y \sin x \cos x - y + 2y^2 e^{xy^2})\, dx + (-x + \sin^2 x + 4xye^{xy^2})\, dy = 0$$

and identify

$$M = 2y \sin x \cos x - y + 2y^2 e^{xy^2} \qquad \text{and} \qquad N = -x + \sin^2 x + 4xye^{xy^2}.$$

Since

$$\frac{\partial M}{\partial y} = 2 \sin x \cos x - 1 + 4xy^3 e^{xy^2} + 4ye^{xy^2} = \frac{\partial N}{\partial x}$$

the equation is exact. Therefore,

$$\frac{\partial f}{\partial x} = 2y \sin x \cos x - y + 2y^2 e^{xy^2}$$

and

$$f = y \sin^2 x - xy + 2e^{xy^2} + g(y).$$

Now

$$\frac{\partial f}{\partial y} = \sin^2 x - x + 4xye^{xy^2} + g'(y) = N = -x + \sin^2 x + 4xye^{xy^2}$$

which implies $g'(y) = 0$ or $g(y) = 0$. Hence, the solution of the differential equation is

$$y \sin^2 x - xy + 2e^{xy^2} = c.$$

28. Write the equation in the form

$$\frac{x}{2y^4}\, dx + \frac{3y^2 - x^2}{y^5}\, dy = 0$$

and identify

$$M = \frac{x}{2y^4} \qquad \text{and} \qquad N = \frac{3y^2 - x^2}{y^5}.$$

Since

$$\frac{\partial M}{\partial y} = -\frac{2x}{y^5} = \frac{\partial N}{\partial x}$$

the equation is exact. Therefore,

$$\frac{\partial f}{\partial x} = \frac{x}{2y^4}$$

and

$$f = \frac{x^2}{4y^4} + g(y).$$

Now

$$\frac{\partial f}{\partial y} = -\frac{x^2}{y^5} + g'(y) = N = \frac{3y^2 - x^2}{y^5}$$

and so

$$g'(y) = \frac{3y^2}{y^5} = 3y^{-3}$$

and

$$g(y) = -\frac{3}{2}y^{-2} = -\frac{3}{2y^2}.$$

Hence, the solution of the differential equation is

$$\frac{x^2}{4y^4} - \frac{3}{2y^2} = c$$

or

$$x^2 - 6y^2 = c_1 y^4.$$

Letting $x = 1$ and $y = 1$ we have $1 - 6 = c_1$, so $c_1 = -5$ and

$$x^2 - 6y^2 = -5y^4.$$

34. We identify

$$M = 6xy^3 + \cos y \qquad \text{and} \qquad N = kx^2y^2 - x\sin y.$$

For the differential equation to be exact we must have $\partial M/\partial y = \partial N/\partial x$ or

$$18xy^2 - \sin y = 2kxy^2 - \sin y.$$

Thus, $k = 9$.

Exercises 2.5

6. Write the equation in the form
$$\frac{dy}{dx} - y = e^x.$$
The integrating factor is $e^{\int(-1)\,dx} = e^{-x}$. Therefore
$$e^{-x}\frac{dy}{dx} - e^{-x}y = 1,$$
which is the same as
$$\frac{d}{dx}[e^{-x}y] = 1.$$
Integrating this last equation then gives
$$e^{-x}y = x + c$$
or
$$y = (x + c)e^x$$
on $(-\infty, \infty)$.

14. Dividing the equation by dx we obtain
$$(1 + x^2)\frac{dy}{dx} + xy = -x^3 - x$$
or
$$\frac{dy}{dx} + \frac{x}{1 + x^2}y = -x.$$
Identifying $P(x) = \dfrac{x}{1 + x^2}$, we see that the integrating factor is
$$e^{\int x\,dx/(1+x^2)} = e^{\frac{1}{2}\ln(1+x^2)} = (1 + x^2)^{1/2}.$$
Thus
$$(1 + x^2)^{1/2}\frac{dy}{dx} + \frac{x}{1 + x^2}(1 + x^2)^{1/2}y = -x(1 + x^2)^{1/2},$$
which is the same as
$$\frac{d}{dx}[(1 + x^2)^{1/2}y] = -x(1 + x^2)^{1/2}.$$
Integrating this last equation gives
$$(1 + x^2)^{1/2}y = -\frac{1}{2}\frac{(1 + x^2)^{3/2}}{3/2} + c$$
or

$$y = -\frac{1}{3}(1+x^2) + c(1+x^2)^{-1/2}$$

on $(-\infty, \infty)$.

20. Write the equation in the form

$$\frac{dy}{dx} - \frac{x}{x+1}y = x.$$

The integrating factor is

$$e^{-\int x\,dx/(x+1)} = e^{-\int[1-1/(x+1)]\,dx} = e^{-x+\ln|x+1|} = e^{-x}(x+1).$$

Therefore

$$e^{-x}(x+1)\frac{dy}{dx} - \frac{x}{x+1}e^{-x}(x+1)\,y = xe^{-x}(x+1),$$

which is the same as

$$\frac{d}{dx}[e^{-x}(x+1)\,y] = e^{-x}(x^2+x).$$

Integrating this last equation gives

$$e^{-x}(x+1)\,y = \int e^{-x}(x^2+x)\,dx.$$

Letting $u = x^2 + x$ and $dv = e^{-x}\,dx$ and using integration by parts, we obtain

$$e^{-x}(x+1)\,y = -e^{-x}(x^2+x) + \int e^{-x}(2x+1)\,dx.$$

Letting $u = 2x+1$ and $dv = e^{-x}dx$ and using integration by parts again, we obtain

$$\begin{aligned} e^{-x}(x+1)\,y &= -e^{-x}(x^2+x) - e^{-x}(2x+1) + 2\int e^{-x}\,dx \\ &= -e^{-x}(x^2+x) - e^{-x}(2x+1) - 2e^{-x} + c, \end{aligned}$$

or

$$y = -x - \frac{2x+3}{x+1} + \frac{ce^x}{x+1}$$

on $(-1, \infty)$.

36. Write the equation in the form

$$\frac{dP}{dt} + (2t-1)P = 2(2t-1).$$

The integrating factor is

$$e^{\int(2t-1)\,dt} = e^{t^2-t}.$$

Thus

$$\frac{d}{dt}\left[e^{t^2-t}P\right] = 2(2t-1)e^{t^2-t}.$$

Integrating this last equation gives

$$e^{t^2-t}P = 2\int (2t-1)e^{t^2-t}\,dt$$
$$= 2e^{t^2-t} + c$$

or

$$P = 2 + ce^{t-t^2}$$

on $(-\infty, \infty)$.

48. Write the equation in the form

$$x\frac{dy}{dx} + (x+2)\,y = 2e^{-x}.$$

Then

$$\frac{dy}{dx} + \frac{x+2}{x}y = 2\frac{e^{-x}}{x},$$

and the integrating factor is

$$e^{\int (x+2)\,dx/x} = e^{\int (1+2/x)\,dx}$$
$$= e^{x+2\ln x}$$
$$= x^2e^x.$$

Thus

$$\frac{d}{dx}[x^2e^xy] = 2\frac{e^{-x}}{x}x^2e^x = 2x.$$

Integrating gives

$$x^2e^xy = \int 2x\,dx = x^2 + c,$$

or

$$y = e^{-x} + \frac{ce^{-x}}{x^2}.$$

Letting $x = 1$ and $y = 0$ in the last equation, we have

$$0 = e^{-1} + ce^{-1}$$

so that $c = -1$. The solution of the initial value problem is

$$y = e^{-x} - \frac{e^{-x}}{x^2}$$

on $(0, \infty)$.

54. Write the equation in the form

$$\frac{dy}{dx} + (\sec^2 x)\,y = \sec^2 x.$$

The integrating factor is

$$e^{\int \sec^2 x\, dx} = e^{\tan x}.$$

Thus

$$\frac{d}{dx}[ye^{\tan x}] = (\sec^2 x)e^{\tan x}.$$

Integrating gives

$$\begin{aligned} ye^{\tan x} &= \int (\sec^2 x)e^{\tan x}\, dx \\ &= e^{\tan x} + c \end{aligned}$$

or

$$y = 1 + ce^{-\tan x}.$$

Letting $x = 0$ and $y = -3$ in the last equation, we have

$$-3 = 1 + c,$$

so $c = -4$. The solution of the initial value problem is

$$y = 1 - 4e^{-\tan x}$$

on $(-\pi/2, \pi/2)$.

Exercises 2.6

6. Write the equation in the form

$$\frac{dy}{dx} + \frac{2x}{3(1+x^2)}y = \frac{2x}{3(1+x^2)}y^4.$$

Then $n = 4$ and the substitution $w = y^{-3}$ leads to the linear equation

$$\frac{dw}{dx} - \frac{2x}{1+x^2}w = -\frac{2x}{1+x^2}.$$

The integrating factor is

$$e^{-\int 2x\, dx/(1+x^2)} = e^{-\ln(1+x^2)} = \frac{1}{1+x^2}.$$

Thus

$$\begin{aligned} \frac{d}{dx}\left[\frac{w}{1+x^2}\right] &= -\frac{2x}{(1+x^2)^2} \\ \frac{w}{1+x^2} &= -\int \frac{2x}{(1+x^2)^2}\, dx \\ &= \frac{1}{1+x^2} + c \end{aligned}$$

and

$$w = 1 + c(1 + x^2).$$

Since $w = y^{-3}$, we have

$$y = \frac{1}{\sqrt[3]{1 + c + cx^2}}.$$

8. Rewrite the given equation as

$$\frac{dy}{dx} + y = y^{-1/2}.$$

Identifying $n = -1/2$ we find that the substitution $w = y^{3/2}$ leads to the linear equation

$$\frac{dw}{dx} + \frac{3}{2}w = \frac{3}{2}.$$

Since the integrating factor for this last equation is $e^{3x/2}$, we obtain

$$\frac{d}{dx}[e^{3x/2}w] = \frac{3}{2}e^{3x/2}$$
$$e^{3x/2}w = e^{3x/2} + c$$
$$w = 1 + ce^{-3x/2}$$
$$y^{3/2} = 1 + ce^{-3x/2}.$$

Now the initial condition $y(0) = 4$ implies $c = 7$. Thus

$$y^{3/2} = 1 + 7e^{-3x/2}.$$

14. First we identify $P(x) = 2x^2$, $Q(x) = 1/x$, and $R(x) = -2$. Using $y_1 = x$ we look for a function u satisfying

$$\frac{du}{dx} - (\frac{1}{x} - 4x)\,u = 2u^2.$$

Identifying $n = 2$, the substitution $w = u^{-1}$ leads to

$$\frac{dw}{dx} + \left(\frac{1}{x} - 4x\right)w = 2.$$

The integrating factor for this equation is

$$e^{\int(1/x - 4x)\,dx} = xe^{-2x^2}.$$

Hence, we have

$$\frac{d}{dx}[xe^{-2x^2}w] = 2xe^{-2x^2}$$
$$xe^{-2x^2}w = -\frac{1}{2}e^{-2x^2} + c$$
$$w = -\frac{1}{2x} + \frac{c}{x}e^{2x^2}.$$

Then

$$u = \frac{1}{w} = \frac{1}{-\frac{1}{2x} + \frac{c}{x}e^{2x^2}}$$

and

$$y = y_1 + u = x + \frac{1}{-\frac{1}{2x} + \frac{c}{x}e^{2x^2}}$$

$$= x + \frac{2x}{-1 + 2ce^{2x^2}}.$$

18. Write the equation as

$$\frac{dy}{dx} = (3 + y)^2.$$

We note that this Ricatti equation is separable. Thus

$$\frac{dy}{(3+y)^2} = dx$$

$$\int \frac{dy}{(3+y)^2} = \int dx$$

$$-\frac{1}{3+y} = x + c$$

$$-\frac{1}{x+c} = 3 + y$$

$$y = -3 - \frac{1}{x+c}.$$

20. With the identification $f(y') = (y')^{-2}$ it follows that a solution of the given Clairant equation is

$$y = cx + c^{-2}.$$

To obtain the singular solution we use

$$x = -f'(t)$$

$$y = f(t) - tf'(t).$$

Since $f(t) = t^{-2}$, $f'(t) = -2t^{-3}$ and so

$$x = 2t^{-3}$$

$$y = t^{-2} - t(-2t^{-3}) = 3t^{-2}.$$

Squaring the first equation and cubing the second, we obtain

$$x^2 = 4t^{-6}$$

$$y^3 = 27t^{-6}.$$

Eliminating the parameter gives the singular solution

$$4y^3 = 27x^2.$$

Exercises 2.7

2. Let $u = \ln y$. Then $\dfrac{du}{dx} = \dfrac{1}{y}\dfrac{dy}{dx}$, and since the given equation can be written as

$$\frac{1}{y}\frac{dy}{dx} + \ln y = e^x,$$

it follows that

$$\frac{du}{dx} + u = e^x.$$

This equation is linear with the integrating factor e^x. thus

$$\frac{d}{dx}[e^x u] = e^{2x}$$

$$e^x u = \frac{1}{2}e^{2x} + c$$

$$u = \frac{1}{2}e^x + ce^{-x}$$

and

$$\ln y = \frac{1}{2}e^x + ce^{-x}.$$

6. Let $u = x + y$. Then $du/dx = 1 + dy/dx$, and the equation becomes

$$\left(\frac{du}{dx} - 1\right) + u + 1 = u^2e^{3x}$$

or

$$\frac{du}{dx} + u = u^2e^{3x}.$$

From Section 2.6 we recognize this as a Bernoulli equation. Using the substitution $w = u^{-1}$ we obtain the linear equation

$$\frac{dw}{dx} - w = -e^{3x}.$$

An integrating factor is e^{-x}, so

$$\frac{d}{dx}[e^{-x}w] = -e^{2x}$$

$$e^{-x}w = -\frac{1}{2}e^{2x} + c$$

$$w = -\frac{1}{2}e^{3x} + ce^{x}$$

$$u^{-1} = -\frac{1}{2}e^{3x} + ce^{x}$$

$$u = \frac{1}{-\frac{1}{2}e^{3x} + ce^{x}}$$

$$x + y = \frac{1}{-\frac{1}{2}e^{3x} + ce^{x}}.$$

The solution is

$$y = \frac{2}{-e^{3x} + c_1e^{x}} - x.$$

22. Letting $w = y'$ the given equation can be written as

$$\frac{dw}{dx} = 1 + w^2.$$

By separation of variables it follows that

$$\frac{dw}{1+w^2} = dx$$

$$\int \frac{dw}{1+w^2} = \int dx$$

$$\tan^{-1} w = x + c_1$$

$$w = \tan(x + c_1)$$

$$\frac{dy}{dx} = \tan(x + c_1)$$

$$dy = \tan(x + c_1)\,dx$$

$$\int dy = \int \tan(x + c_1)\,dx$$

$$y = -\ln|\cos(x + c_1)| + c_2.$$

$$= \ln|\sec(x + c_1)| + c_2.$$

Exercises 2.8

6. Identify $x_0 = 0$, $y_0 = 1$, and $f(t, y_{n-1}(t)) = 2e^t - y_{n-1}(t)$. Then Picard's iteration formula is

$$y_n(x) = 1 + \int_0^x (2e^t - y_{n-1}(t))\,dt, \quad n = 1, 2, 3, \ldots.$$

Thus

$$\begin{aligned}
y_1(x) &= 1 + \int_0^x (2e^t - 1)\,dt \\
&= 1 + \Big[2e^t - t\Big]_0^x \\
&= 1 + 2e^x - x - 2 \\
&= -1 + 2e^x - x \\
y_2(x) &= 1 + \int_0^x [2e^t - (-1 + 2e^t - t)]\,dt \\
&= 1 + \int_0^x (1 + t)\,dt \\
&= 1 + \left[t + \frac{t^2}{2}\right]_0^x \\
&= 1 + x + \frac{x^2}{2} \quad . \\
y_3(x) &= 1 + \int_0^x \left[2e^t - \left(1 + t + \frac{t^2}{2}\right)\right] dt \\
&= 1 + \left[2e^t - t - \frac{t^2}{2} - \frac{t^3}{6}\right]_0^x \\
&= 1 + 2e^x - x - \frac{x^2}{2} - \frac{x^3}{6} - 2 \\
&= -1 + 2e^x - x - \frac{x^2}{2} - \frac{x^3}{6} \\
y_4(x) &= 1 + \int_0^x \left[2e^t - \left(-1 + 2e^t - t - \frac{t^2}{2} - \frac{t^3}{6}\right)\right] dt \\
&= 1 + \left[2e^t + t - 2e^t + \frac{t^2}{2} + \frac{t^3}{6} + \frac{t^4}{24}\right]_0^x \\
&= 1 + x + \frac{x^2}{2} + \frac{x^3}{6} + \frac{x^4}{24}.
\end{aligned}$$

For n an integer we conclude that

$$\lim_{n\to\infty} y_{2n} = \sum_{k=0}^{\infty} \frac{x^k}{k!} = e^x$$

and

$$\begin{aligned}\lim_{n\to\infty} y_{2n+1} &= 2e^x - \sum_{k=0}^{\infty} \frac{x^k}{k!} \\ &= 2e^x - e^x \\ &= e^x.\end{aligned}$$

The solution of the initial value problem is $y = e^x$.

3 Applications of First-Order Differential Equations

Exercises 3.1

8. Differentiating y and using $c_1 = (\ln y)/x$ we obtain

$$\begin{aligned}
\frac{dy}{dx} &= c_1 e^{c_1 x} \\
\frac{dy}{dx} &= c_1 y \\
\frac{dy}{dx} &= \frac{y \ln y}{x}.
\end{aligned}$$

The differential equation of the orthogonal family is

$$\frac{dy}{dx} = -\frac{x}{y \ln y}.$$

Separating variables and using integration by parts, we have

$$\begin{aligned}
y \ln y \, dy &= -x \, dx \\
\int y \ln y \, dy &= -\int x \, dx \\
\frac{y^2}{2} \ln y - \frac{y^2}{4} &= -\frac{x^2}{2} + c_2 \\
2y^2 \ln y - y^2 &= -2x^2 + c_3.
\end{aligned}$$

14. Differentiating the equation and solving for dy/dx we obtain

$$\begin{aligned}
2x + 2y\frac{dy}{dx} &= 2c_1 \\
2x + 2y\frac{dy}{dx} &= \frac{x^2 + y^2}{x} \\
2xy\frac{dy}{dx} &= y^2 - x^2 \\
\frac{dy}{dx} &= \frac{y^2 - x^2}{2xy}.
\end{aligned}$$

The differential equation of the orthogonal family is

$$\frac{dy}{dx} = \frac{-2xy}{y^2 - x^2}$$

or

$$(y^2 - x^2)\, dy + 2xy\, dx = 0.$$

Since this equation is homogeneous, the substitution $x = vy$ reduces the equation to

$$\frac{dy}{y} + \frac{2v\, dv}{1 + v^2} = 0.$$

Integrating we find

$$\ln|y| + \ln(1 + v^2) = \ln c_2$$

$$y(1 + v^2) = c_3$$

$$y\left(1 + \frac{x^2}{y^2}\right) = c_3$$

$$x^2 + y^2 = c_3 y.$$

20. Differentiating and eliminating c_1 we obtain

$$\frac{dy}{dx} = -1 + c_1 e^x$$

$$\frac{dy}{dx} = -1 + (y + x + 1)$$

$$= y + x.$$

The differential equation of the orthogonal family is

$$\frac{dy}{dx} = \frac{-1}{y + x}$$

or

$$\frac{dx}{dy} = -x - y.$$

In the form

$$\frac{dx}{dy} + x = -y$$

we recognize the equation as linear in x. The integrating factor is e^y and thus

$$\frac{d}{dy}[e^y x] = -ye^y$$

$$e^y x = -\int ye^y\, dy$$

$$= -ye^y + e^y + c_2$$

or

$$x = -y + 1 + c_2e^{-y}.$$

28. Differentiating and eliminating c_1 we obtain

$$\begin{aligned}
6xy\,\frac{dy}{dx} + 3y^2 &= 3c_1 \\
6xy\,\frac{dy}{dx} + 3y^2 &= \frac{3xy^2 - 2}{x} \\
6x^2y\,\frac{dy}{dx} + 3xy^2 &= 3xy^2 - 2 \\
6x^2y\,\frac{dy}{dx} &= -2 \\
\frac{dy}{dx} &= -\frac{1}{3x^2y}\,.
\end{aligned}$$

The differential equation of the orthogonal family is

$$\frac{dy}{dx} = 3x^2y$$

or

$$\frac{dy}{y} = 3x^2\,dx.$$

Thus

$$\begin{aligned}
\int \frac{dy}{y} &= 3\int x^2\,dx \\
\ln y &= x^3 + c_2 \\
y &= c_3e^{x^3}.
\end{aligned}$$

Substituting $x = 0$ and $y = 10$ into the last equation, we find $c_3 = 10$. The orthogonal trajectory through $(0, 10)$ is

$$y = 10e^{x^3}.$$

32. Differentiating and eliminating c_1 we obtain

$$\frac{dr}{d\theta} = \frac{c_1\sin\theta}{(1+\cos\theta)^2} = \frac{r\sin\theta}{1+\cos\theta}\,.$$

Then

$$r\frac{d\theta}{dr} = \frac{1+\cos\theta}{\sin\theta} = \tan\psi_1$$

and the differential equation of the orthogonal trajectories is

$$r\frac{d\theta}{dr} = -\frac{\sin\theta}{1+\cos\theta} = \tan\psi_2.$$

Separating variables we have

$$\begin{aligned}\frac{dr}{r} &= -\frac{1+\cos\theta}{\sin\theta}\,d\theta \\ &= -(\csc\theta + \cot\theta)\,d\theta.\end{aligned}$$

Integrating gives

$$\begin{aligned}\ln|r| &= -\ln|\csc\theta - \cot\theta| - \ln|\sin\theta| + \ln c_2 \\ &= \ln\frac{c_2}{|1-\cos\theta|}.\end{aligned}$$

Hence

$$r = \frac{c_2}{1-\cos\theta}.$$

38. Since $\alpha = 30°$, $\tan\alpha = \sqrt{3}/3$. The differential equation of the family $y = c_1 x$ is

$$\frac{dy}{dx} = c_1 = \frac{y}{x} = f(x,y).$$

By Problem 35 the differential equation of the isogonal family is

$$\frac{dy}{dx} = \frac{y/x \pm \sqrt{3}/3}{1 \mp (y/x)\sqrt{3}/3}.$$

Make the substitution $u = y/x$. Then $dy/dx = u + x\,du/dx$, and

$$\begin{aligned}u + x\frac{du}{dx} &= \frac{u \pm \sqrt{3}/3}{1 \mp u\sqrt{3}/3} \\ x\frac{du}{dx} &= \frac{3u \pm \sqrt{3}}{3 \mp \sqrt{3}\,u} - u \\ x\frac{du}{dx} &= \frac{3u \pm \sqrt{3} - 3u + \sqrt{3}\,u^2}{3 \mp \sqrt{3}\,u} \\ x\frac{du}{dx} &= \pm\sqrt{3}\left(\frac{1+u^2}{3 \mp \sqrt{3}\,u}\right).\end{aligned}$$

This is a separable differential equation and

$$\begin{aligned}\pm\left(\frac{3 \mp \sqrt{3}\,u}{1+u^2}\right) du &= \sqrt{3}\,\frac{dx}{x} \\ \int\left(\frac{\pm 3 - \sqrt{3}\,u}{1+u^2}\right) du &= \sqrt{3}\int\frac{dx}{x} \\ \pm 3\tan^{-1}u - \frac{\sqrt{3}}{2}\ln(1+u^2) &= \sqrt{3}\,\ln|x| + \sqrt{3}\,\ln c_2 \\ \pm 6\tan^{-1}\frac{y}{x} - \sqrt{3}\,\ln\left(1+\frac{y^2}{x^2}\right) &= 2\sqrt{3}\,\ln c_2 x.\end{aligned}$$

Simplifying this equation we obtain the equation of the family of isogonal trajectories

$$\begin{aligned}\pm 6\tan^{-1}\frac{y}{x} &= \sqrt{3}\left[\ln c_2^2x^2+\ln\left(1+\frac{y^2}{x^2}\right)\right]\\ \tan^{-1}\frac{y}{x} &= \pm\frac{\sqrt{3}}{6}\ln c_3(x^2+y^2)\\ \frac{y}{x} &= \tan\left[\pm\frac{\sqrt{3}}{6}\ln c_3(x^2+y^2)\right]\\ \frac{y}{x} &= \pm\tan\left[\frac{\sqrt{3}}{6}\ln c_3(x^2+y^2)\right]\\ y &= \pm x\tan\left[\frac{\sqrt{3}}{6}\ln c_3(x^2+y^2)\right].\end{aligned}$$

Exercises 3.2

2. From Problem 1 we know that

$$P(t) = P_0e^{0.1386t}.$$

Since $P(3) = 10,000$, solving

$$10,000 = P_0e^{(0.1386)3}$$

for P_0 gives

$$P_0 = 10,000e^{-0.4158} \approx 6598.$$

Thus, in 10 years,

$$P(10) = 6589e^{(0.1386)10} \approx 26,348.$$

8. The solution of $dA/dt = kA$ is

$$A(t) = A_0e^{kt},$$

so

$$\begin{aligned}A_1 &= A(t_1) = A_0e^{kt_1}\\ A_2 &= A(t_2) = A_0e^{kt_2}\\ \frac{A_1}{A_2} &= e^{k(t_1-t_2)}\\ k(t_1-t_2) &= \ln\frac{A_1}{A_2}\\ k &= \frac{1}{(t_1-t_2)}\ln\frac{A_1}{A_2}.\end{aligned}$$

Solving $A_0/2 = A_0 e^{kt}$ for t, we obtain $t = -(\ln 2)/k$. It follows that

$$t = \frac{(t_2 - t_1)\ln 2}{\ln(A_1/A_2)}.$$

12. Given that $T_0 = 5$, the solution of

$$\frac{dT}{dt} = k(T - 5)$$

is

$$T(t) = 5 + ce^{kt}.$$

When $t = 1$,

$$55 = 5 + ce^{k}$$

or

$$ce^{k} = 50. \tag{1}$$

When $t = 5$,

$$30 = 5 + ce^{5k}$$

or

$$ce^{5k} = 25. \tag{2}$$

Dividing equation (2) by equation (1) gives

$$e^{4k} = \frac{1}{2},$$

so that

$$\begin{aligned} 4k &= -\ln 2 \\ k &= -\frac{1}{4}\ln 2 \\ &\approx -0.1733. \end{aligned}$$

From (1) we find that

$$\begin{aligned} c &= 50e^{-k} \\ &\approx 50e^{0.1733} \\ &\approx 59.4611. \end{aligned}$$

Therefore,

$$T(t) = 5 + 59.4611e^{-0.1733t}.$$

Now at $t = 0$ we find the initial temperature to be

$$T(0) \approx 64.4611^{\circ}.$$

18. The linear differential equation

$$1000\frac{dq}{dt} + \frac{1}{5}10^6 q = 200$$

can be written as

$$\frac{dq}{dt} + 200q = \frac{1}{5},$$

and has the solution

$$q(t) = \frac{1}{1000} + ce^{-200t}.$$

Now

$$i(t) = \frac{dq}{dt} = -200ce^{-200t}$$

and $i(0) = 0.4$ implies $c_1 = -1/500$. Therefore

$$q(t) = \frac{1}{1000} - \frac{1}{500}e^{-200t}$$

and

$$i(t) = \frac{2}{5}e^{-200t}.$$

It follows that

$$q(0.005) = 0.0003 \text{ coulombs} \qquad \text{and} \qquad i(0.005) = 0.1472 \text{ amperes}.$$

Also, $q(t) \to 1/1000$ as $t \to \infty$.

22. If $A(t)$ is the amount of salt in the tank at any time, then

$$\frac{dA}{dt} = (\text{rate of substance entering}) - (\text{rate of substance leaving})$$
$$= R_1 - R_2.$$

In liters per minute, the rate at which salt is entering is

$$R_1 = (4 \text{ l/min}) \cdot (0 \text{ g/l})$$
$$= 0 \text{ g/min}.$$

But the rate at which salt leaves the tank is

$$R_2 = (4 \text{ l/min}) \cdot (\frac{A}{200} \text{ g/l})$$
$$= \frac{A}{50} \text{ g/min}.$$

Thus

$$\frac{dA}{dt} = -\frac{A}{50}.$$

By separating variables we find that

$$A = ce^{-t/50}.$$

Since $A(0) = 30$ g we have $c = 30$. Hence,

$$A(t) = 30e^{-t/50}.$$

30. We express Fick's law as

$$\frac{dm}{dt} = kA[\,C_s - C(t)\,].$$

Since $m = VC$

$$\frac{dm}{dt} = V\frac{dC}{dt} = kA[\,C_s - C(t)\,]$$

and

$$\frac{dC}{dt} + \frac{kA}{V}C = \frac{kA}{V}C_s.$$

This is a linear equation whose solution is

$$C(t) = C_s + c_1e^{-kAt/V}.$$

Applying the initial condition $C(0) = C_0$ then yields $c_1 = C_0 - C_s$. Hence we have

$$C(t) = C_s + (C_0 - C_s)e^{-kAt/V}.$$

Exercises 3.3

2. From equation (6) in the text we see that the initial value problem

$$\frac{dN}{dt} = N(a - bN), \quad N(0) = 500$$

has the solution

$$N(t) = \frac{500a}{500b + (a - 500b)e^{-at}}.$$

Now $\lim_{t\to\infty} N(t) = a/b$ and so $a/b = 50,000$ or $b = a/50,000$. Therefore, $N(t)$ can be written entirely in terms of a:

$$N(t) = \frac{500a}{\frac{a}{100} + (a - \frac{a}{100})e^{-at}} = \frac{50,000}{1 + 99e^{-at}}.$$

But, $N(1) = 1000$ implies

$$1000 = \frac{50,000}{1 + 99e^{-a}}$$
$$1 + 99e^{-a} = 50$$
$$e^{-a} = \frac{49}{99}$$

and

$$a \approx 0.7033.$$

Therefore

$$N(t) = \frac{50,000}{1 + 99e^{-0.7033t}}.$$

8. Referring to the law of mass action we identify $M = 2$, $N = 1$, $a = 100$, and $b = 50$. Then $\alpha = 150$ and $\beta = 150$. From equation (15) in the text we have the differential equation

$$\frac{dX}{dt} = k(150 - X)^2.$$

Using separation of variables,

$$\int \frac{dX}{(150 - X)^2} = \int k\,dt$$
$$\frac{1}{150 - X} = kt + c_1$$
$$X = 150 - \frac{1}{kt + c_1}$$

When $t = 0$, $X = 0$, so $c_1 = 1/150$; and when $t = 5$, $X = 10$, so

$$10 = 150 - \frac{1}{5k + \frac{1}{150}}$$

or $k = 1/10500$. Thus,

$$X = 150 - \frac{1}{\frac{t}{10500} + \frac{1}{150}}$$
$$= 150 - \frac{10500}{t + 70}$$

For $t = 20$ minutes, $X = 33.33$ grams. The limiting amount of C is 150 grams. Since the original amounts of A and B were 100 and 50 grams, respectively, and we have just seen that the limiting amount of C is 150 grams, we conclude that A and B are completely used up after a long time. To determine when C is half-formed we solve

$$75 = 150 - \frac{10500}{t + 70}$$

for t. This gives $t = 70$ minutes.

14. Separating variables in the given equation:

$$\frac{dT}{T^4 - T_0^4} = k\,dt$$

$$\int \frac{dT}{(T+T_0)(T-T_0)(T^2+T_0^2)} = \int k\,dt.$$

By partial fractions this last equation can be written as

$$\int \left[\frac{1/(4T_0^3)}{T-T_0} - \frac{1/(4T_0^3)}{T+T_0} - \frac{1/(2T_0^2)}{T^2+T_0^2} \right] dt = \int k\,dt.$$

it follows that

$$\frac{1}{4T_0^3}\ln|T-T_0| - \frac{1}{4T_0^3}\ln|T+T_0| - \frac{1}{2T_0^3}\tan^{-1}\frac{T}{T_0} = kt + c.$$

Simplifying this last expression gives

$$\ln\left|\frac{T-T_0}{T+T_0}\right| - 2\tan^{-1}\frac{T}{T_0} = 4T_0^3 kt + c_1.$$

16. When $h = 0$ the given equation can be written as

$$r^{1/2}dr = \sqrt{2\mu}\,dt.$$

Integrating we obtain

$$\frac{2}{3}r^{3/2} = \sqrt{2\mu}\,t + c_1, \quad \mu > 0.$$

For $h > 0$, the equation separates as

$$\left(\frac{2\mu}{r} + 2h\right)^{-1/2} dr = dt$$

or

$$\frac{r^{1/2}}{\sqrt{2\mu + 2hr}}\,dr = dt.$$

With the aid of a table of integrals, it follows that

$$\frac{1}{2h}(2\mu r + 2hr^2)^{1/2} - \frac{2\mu}{(2h)^{3/2}}\ln(\sqrt{\mu + hr} + \sqrt{hr}) = t + c_2.$$

20. Writing the equation as

$$x\left(\frac{dx}{dy}\right)^2 + 2y\left(\frac{dx}{dy}\right) - x = 0$$

we recognize a quadratic equation in dx/dy. It folows from the quadratic formula that

$$\frac{dx}{dy} = \frac{-y \pm \sqrt{x^2 + y^2}}{x}.$$

Although this last equation is homogeneous, it can more easily be solved by writing the equation in the form

$$\frac{x\,dx + y\,dy}{\sqrt{x^2 + y^2}} = \pm dy,$$

and noting that the left side is the exact differential $d(\sqrt{x^2 + y^2})$. Then

$$\sqrt{x^2 + y^2} = \pm y + c$$

or

$$x^2 = \pm 2cy + c^2.$$

4 Linear Differential Equations of Higher Order

Exercises 4.1

4. From

$$y = c_1 + c_2 \cos x + c_3 \sin x$$

we find

$$y' = -c_2 \sin x + c_3 \cos x$$

and

$$y'' = -c_2 \cos x - c_3 \sin x.$$

Then

$$0 = y(\pi) = c_1 + c_2 \cos \pi + c_3 \sin \pi = c_1 - c_2$$

$$2 = y'(\pi) = -c_2 \sin \pi + c_3 \cos \pi = -c_3$$

$$-1 = y''(\pi) = -c_2 \cos \pi - c_3 \sin \pi = c_2$$

so that $c_1 = 1$, $c_2 = -1$, and $c_3 = -2$. The solution is

$$y = -1 - \cos x - 2 \sin x.$$

8. From

$$y = c_1 + c_2 x^2$$

we find

$$y' = 2c_2 x.$$

Then $1 = y(0) = c_1$ and $6 = y'(1) = 2c_2$, so that $c_1 = 1$ and $c_2 = 3$. The solution is $y = 1 + 3x^2$. Theorem 4.1 does not apply because y and y' are evaluated at different points. Also, $a_n(x) = x$ is zero for $x = 0$.

14. By Problem 13, a general solution of the equation $y'' + \lambda^2 y = 0$ is

$$y = c_1 \cos \lambda x + c_2 \sin \lambda x.$$

Now, applying the first boundary condition gives

$$0 = y(0) = c_1 \cdot 1 + c_2 \cdot 0$$

or $c_1 = 0$. Thus,

$$y = c_2 \sin \lambda x.$$

The second condition implies

$$0 = y(5) = c_2 \sin 5\lambda.$$

If $c_2 = 0$ then we get $y \equiv 0$. However, if

$$\sin 5\lambda = 0$$

then $5\lambda = n\pi$ or $\lambda = n\pi/5$ for $n = 1, 2, 3, \ldots$. Hence, the one-parameter family

$$y = c_2 \sin \frac{n\pi}{5} x$$

is a solution to the problem

$$y'' + \left(\frac{n\pi}{5}\right)^2 y = 0, \quad y(0) = 0, \quad y(5) = 0$$

for any positive integer choice of n.

16. Since

$$1 \cdot 0 + 0 \cdot x + 0 \cdot e^x = 0$$

we see that 0, x, and e^x are linearly dependent. A similar argument shows that any set of functions containing $f(x) = 0$ will be linearly dependent.

30. (a) The graphs of f_1 and f_2 are as follows:

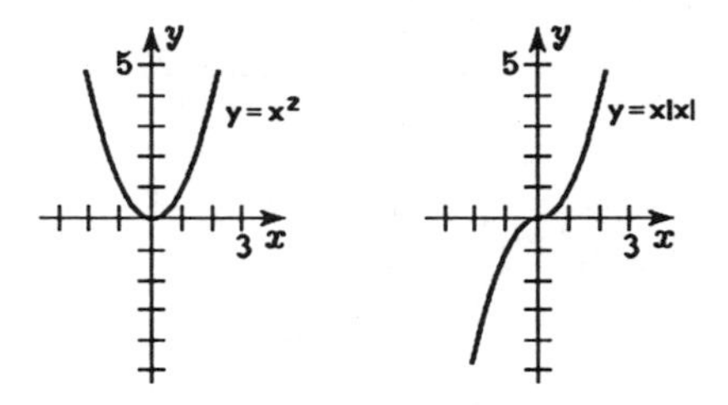

Obviously, neither function is a constant multiple of the other on $-\infty < x < \infty$. Hence, f_1 and f_2 are linearly independent on $(-\infty, \infty)$.

(b) For $x \geq 0$, $f_2 = x^2$ and so

$$W(f_1, f_2) = \begin{vmatrix} x^2 & x^2 \\ 2x & 2x \end{vmatrix} = 2x^3 - 2x^3 = 0.$$

For $x < 0$, $f_2 = -x^2$ and

$$W(f_1, f_2) = \begin{vmatrix} x^2 & -x^2 \\ 2x & -2x \end{vmatrix} = -2x^3 + 2x^3 = 0.$$

We conclude that $W(f_1, f_2) = 0$ for all real values of x.

32. (a) For $y_1 = 1$ we have $y_1' = y_1'' = 0$ and

$$0 + 0^2 = 0$$

so y_1 is a solution of $y'' + (y')^2 = 0$. For $y_2 = \ln x$ we have $y_2' = 1/x$ and $y_2'' = -1/x^2$. Then

$$-\frac{1}{x^2} + \left(\frac{1}{x}\right)^2 = 0,$$

and y_2 is a solution for $x > 0$.

(b) For $y = 1 + \ln x$ we have $y' = 1/x$ and $y'' = -1/x^2$. Then

$$y'' + (y')^2 = -\frac{1}{x^2} + \left(\frac{1}{x}\right)^2 = 0$$

and y is a solution. However, for $y = 1 + 2\ln x$ we have $y' = 2/x$ and $y'' = -2/x^2$ so that

$$y'' + (y')^2 = -\frac{2}{x^2} + \left(\frac{2}{x}\right)^2 = \frac{2}{x^2} \neq 0.$$

Thus, $c_1y_1 + c_2y_2$ is not a solution of the equation for every choice of c_1 and c_2.

38. For $y_1 = \cos(\ln x)$ we have

$$y_1' = -\frac{1}{x}\sin(\ln x)$$

and

$$y_1'' = -\frac{1}{x^2}\cos(\ln x) + \frac{1}{x^2}\sin(\ln x).$$

Then

$$x^2y'' + xy' + y = -\cos(\ln x) + \sin(\ln x) - \sin(\ln x) + \cos(\ln x) = 0$$

and y_1 is a solution. Similarly, for $y_2 = \sin(\ln x)$ we have

$$y_2' = \frac{1}{x}\cos(\ln x)$$

and

$$y_2'' = -\frac{1}{x^2}\sin(\ln x) - \frac{1}{x^2}\cos(\ln x).$$

Then

$$x^2y'' + xy' + y = -\sin(\ln x) - \cos(\ln x) + \cos(\ln x) + \sin(\ln x) = 0$$

and y_2 is a solution. Now

$$W(y_1, y_2) = \begin{vmatrix} \cos(\ln x) & \sin(\ln x) \\ -\dfrac{1}{x}\sin(\ln x) & \dfrac{1}{x}\cos(\ln x) \end{vmatrix}$$

$$= \frac{1}{x}\cos^2(\ln x) + \frac{1}{x}\sin^2(\ln x)$$
$$= \frac{1}{x}$$
$$\neq 0.$$

Thus, y_1 and y_2 are linearly independent, and they form a fundamental set of solutions. The general solution is

$$y = c_1 \cos(\ln x) + c_2 \sin(\ln x).$$

48. We identify $a_2(x) = 1 - x^2$ and $a_1(x) = -2x$. Then from Abel's formula in Problem 47 we have

$$W = ce^{-\int [a_1(x)/a_2(x)]\, dx}$$
$$= ce^{-\int [-2x/(1-x^2)]\, dx}$$
$$= ce^{-\ln(1-x^2)}$$
$$= \frac{c}{1-x^2}.$$

Exercises 4.2

14. Writing the equation in the form

$$y'' + \frac{1}{4x^2}y = 0$$

we make the identification $P(x) = 0$. Therefore, $e^{-\int P(x)\,dx} = e^{\int 0\, dx} = e^0 = 1$, and

$$y_2 = y_1(x)\int \frac{e^{-\int P(x)\,dx}}{y_1^2(x)}\,dx$$
$$= x^{1/2}\ln x \int \frac{dx}{x(\ln x)^2}$$
$$= x^{1/2}\ln x \int (\ln x)^{-2}\frac{dx}{x}$$
$$= x^{1/2}\ln x \cdot \frac{(\ln x)^{-1}}{-1}$$
$$= -x^{1/2}.$$

Since the given equation is homogenous, any real multiple of a solution is also a solution. Thus we can take $y_2 = x^{1/2}$ as a second linearly independent solution.

20. Writing the equation in the form

$$y'' + \frac{x}{1+x}y' - \frac{1}{1+x}y = 0$$

we identify

$$P(x) = \frac{x}{1+x} = 1 - \frac{1}{1+x}.$$

Then

$$\int P(x)\,dx = x - \ln(1+x)$$

and

$$e^{-\int P(x)\,dx} = e^{-x+\ln(1+x)}$$
$$= (1+x)e^{-x}.$$

Thus,

$$y_2 = x\int \frac{(1+x)e^{-x}}{x^2}\,dx$$
$$= x\int\left(\frac{e^{-x}}{x^2} + \frac{e^{-x}}{x}\right)dx$$
$$= x\int \frac{e^{-x}}{x^2}\,dx + x\int \frac{e^{-x}}{x}\,dx,$$

and using integration by parts,

$$y_2 = x\left[-x^{-1}e^{-x} - \int \frac{e^{-x}}{x}\,dx\right] + x\int \frac{e^{-x}}{x}\,dx$$
$$= -e^{-x}.$$

Since the given equation is homogeneous, any real multiple of a solution is also a solution. Thus, we can take $y_2 = e^{-x}$.

24. Writing the equation in the form

$$y'' + \frac{1}{x}y' + \frac{1}{x^2}y = 0$$

we identify $P(x) = 1/x$. Then

$$P(x)\,dx = \ln x$$

and

$$e^{-\int P(x)\,dx} = e^{-\ln x} = \frac{1}{x}.$$

Thus,

$$y_2 = \cos(\ln x)\int \frac{1/x}{\cos^2(\ln x)}\,dx$$

$$= \cos(\ln x) \int \sec^2(\ln x)\,\frac{dx}{x}$$
$$= \cos(\ln x)\tan(\ln x)$$
$$= \sin(\ln x).$$

34. Suppose $y = ue^x$ is a solution of the related homogeneous differential equation. Then

$$y' = e^x u + e^x u'$$
$$y'' = e^x u'' + 2e^x u' + e^x u$$

and

$$y'' - 4y' + 3y = e^x(u'' + 2u' + u) - 4e^x(u' + u) + 3e^x u$$
$$= e^x(u'' - 2u') = 0.$$

Letting $w = u'$ we obtain the first order differential equation $w' - 2w = 0$. Separating variables we have

$$\frac{dw}{w} = 2\,dx$$
$$\ln w = 2x$$
$$w = e^{2x}.$$

Then $u' = e^{2x}$ and $u = \frac{1}{2}e^{2x}$. Since we are now considering a homogeneous differential equation, we may drop the 1/2 and write a second solution of the homogeneous differential equation as $y_2 = ue^x = e^{2x}e^x = e^{3x}$. To find a particular solution we try $y_p = ax + b$. Then

$$y' = a, \qquad y'' = 0,$$

and

$$x = y'' - 4y' + 3y$$
$$= 0 - 4a + 3(ax + b)$$
$$= 3ax + (3b - 4a).$$

We see that $3a = 1$ so that $a = 1/3$, and $3b - 4a = 0$ so that $b = 4/9$. Therefore, a particular solution of the differential is $y_p = \frac{1}{3}x + \frac{4}{9}$.

Exercises 4.3

16. Solving the auxiliary equation $2m^2 - 3m + 4 = 0$ we obtain

$$\begin{aligned} m &= \frac{3 \pm \sqrt{9-32}}{4} \\ &= \frac{3}{4} \pm \frac{\sqrt{23}}{4}\,i. \end{aligned}$$

Thus,

$$y = e^{3x/4}\left[c_1 \cos \frac{\sqrt{23}}{4}\,x + c_2 \sin \frac{\sqrt{23}}{4}\,x\right]$$

26. Observing that $m = 2$ is a root of the auxiliary equation $m^3 - m^2 - 4 = 0$ we divide and obtain $(m-2)(m^2+m+2) = 0$. The roots of this equation are $m = 2,\ -\frac{1}{2} \pm \frac{\sqrt{7}}{2}\,i$. Thus

$$y = c_1 e^{2x} + e^{-x/2}\left[c_2 \cos \frac{\sqrt{7}}{2}\,x + c_3 \sin \frac{\sqrt{7}}{2}\,x\right].$$

32. Factoring the auxiliary equation, we obtain

$$\begin{aligned} m^4 - 7m^2 - 18 &= (m^2-9)(m^2+2) \\ &= (m-3)(m+3)(m^2+2) = 0. \end{aligned}$$

Thus $m = \pm 3,\ \pm\sqrt{2}\,i$ and

$$y = c_1 e^{-3x} + c_2 e^{3x} + c_3 \cos \sqrt{2}\,x + c_4 \sin \sqrt{2}\,x.$$

36. Possible rational roots of the auxiliary equation

$$2m^5 - 7m^4 + 12m^3 + 8m^2 = m^2(2m^3 - 7m^2 + 12m + 8) = 0$$

are 0, ± 1, ± 2, ± 4, ± 8, and $\pm 1/2$. After an exhaustive search we find that $m = -1/2$ is a root, so that $2m+1$ is a factor. Thus, the auxiliary equation may be written in the form

$$m^2(2m+1)(m^2-4m+8) = 0.$$

The roots are 0, 0, $-1/2$, and $2 \pm 2i$. Thus

$$y = c_1 + c_2 x + c_3 e^{-x/2} + e^{2x}[c_4 \cos 2x + c_5 \sin 2x].$$

44. Factoring the auxiliary equation, we obtain

$$4m^2 - 4m - 3 = (2m-3)(2m+1) = 0.$$

Thus $m = -1/2$ and $3/2$, and

$$y = c_1 e^{-x/2} + c_2 e^{3x/2}.$$

To solve the initial value problem we compute

$$y' = -\frac{c_1}{2} e^{-x/2} + \frac{3c_2}{2} e^{3x/2}.$$

Using $y(0) = 1$ and $y'(0) = 5$ we obtain

$$\begin{aligned} c_1 + c_2 &= 1 \\ -\frac{c_1}{2} + \frac{3c_2}{2} &= 5 \end{aligned}$$

or

$$\begin{aligned} c_1 + c_2 &= 1 \\ -c_1 + 3c_2 &= 10. \end{aligned}$$

Adding, we find $4c_2 = 11$ or $c_2 = 11/4$. Then $c_1 = -7/4$, and the solution to the initial value problem is

$$y = -\frac{7}{4} e^{-x/2} + \frac{11}{4} e^{3x/2}.$$

52. Factoring the auxiliary equation, we obtain

$$m^4 - 1 = (m^2 - 1)(m^2 + 1) = (m - 1)(m + 1)(m^2 + 1) = 0.$$

The roots are $m = \pm 1$ and $\pm i$. Thus

$$\begin{aligned} y &= c_1 e^{-x} + c_2 e^{x} + c_3 \cos x + c_4 \sin x \\ y' &= -c_1 e^{-x} + c_2 e^{x} - c_3 \sin x + c_4 \cos x \\ y'' &= c_1 e^{-x} + c_2 e^{x} - c_3 \cos x - c_4 \sin x \\ y''' &= -c_1 e^{-x} + c_2 e^{x} + c_3 \sin x - c_4 \cos x. \end{aligned}$$

Using $y(0) = y'(0) = y''(0) = 0$ and $y'''(0) = 1$ we obtain

$$\begin{aligned} c_1 + c_2 + c_3 \qquad &= 0 \\ -c_1 + c_2 \qquad + c_4 &= 0 \\ c_1 + c_2 - c_3 \qquad &= 0 \\ -c_1 + c_2 \qquad - c_4 &= 1. \end{aligned}$$

Solving, we find $c_1 = -1/4$, $c_2 = 1/4$, $c_3 = 0$, and $c_4 = -1/2$. The solution to the initial value problem is

$$y = -\frac{1}{4}e^{-x} + \frac{1}{4}e^{x} - \frac{1}{2}\sin x.$$

56. The roots of the auxiliary equation $m^2 - 1 = 0$ are 1 and -1, so

$$y = c_1e^x + c_2e^{-x}$$

or

$$y = c_1\cosh x + c_2\sinh x$$

and

$$y' = c_1\sinh x + c_2\cosh x.$$

Using $y(0) = 1$ and $y'(1) = 0$ we obtain $c_1 = 1$ and

$$\sinh 1 + c_2\cosh 1 = 0.$$

Thus, $c_2 = -\sinh 1/\cosh 1$ and the solution to the boundary value problem is

$$\begin{aligned} y &= \cosh x - \frac{\sinh 1}{\cosh 1}\sinh x \\ &= \frac{\cosh x\cosh 1 - \sinh x\sinh 1}{\cosh 1} \\ &= \frac{\cosh(x-1)}{\cosh 1}. \end{aligned}$$

58. Multiplying out the auxiliary equation

$$\begin{aligned} \left(m - \left(-\frac{1}{2}\right)\right)(m-(3+i))(m-(3-i)) &= 0 \\ (2m+1)(m^2-6m+10) &= 0 \\ 2m^3 - 11m^2 + 14m + 10 &= 0, \end{aligned}$$

we see that the corresponding differential equation is

$$2y''' - 11y'' + 14y' + 10y = 0.$$

60. From the solution $y_1 = e^{-4x}\cos x$ we conclude that $m_1 = -4 + i$ and $m_2 = -4 - i$ are roots of the auxiliary equation. Hence another solution must be $y_2 = e^{-4x}\sin x$. Now dividing the polynomial $m^3 + 6m^2 + m - 34$ by

$$[m - (-4+i)][m - (-4-i)] = m^2 + 8m + 17$$

gives $m-2$. Therefore $m_3=2$ is the third root of the auxiliary equation, and the general solution of the differential equation is

$$y = c_1e^{-4x}\cos x + c_2e^{-4x}\sin x + c_3e^{2x}.$$

62. From the solutions $y_1 = 10\cos 4x$ and $y_2 = -5\sin 4x$ we see that $m_1 = 4i$ and $m_2 = -4i$ are roots of the auxiliary equation. This equation is then

$$(m-4i)(m-(-4i)) = m^2+16 = 0.$$

The corresponding differential equation is

$$y'' + 16y = 0.$$

Exercises 4.4

10. The differential operator factors as

$$D^4 - 8D^2 + 16 = (D^2-4)^2 = (D-2)^2(D+2)^2.$$

22. The function $e^{-x}\sin x$ is annihilated by D^2+2D+2 and $e^{2x}\cos x$ is annihilated by D^2-4D+5. Thus, the given function is annihilated by $(D^2-4D+5)(D^2+2D+2)$.

24. From the auxiliary equation

$$2m^2 - 7m + 5 = (2m-5)(m-1) = 0$$

we find the complimentary function

$$y_c = c_1e^{5x/2} + c_2e^x.$$

An annihilator for -29 is D. Applying D to both sides of

$$2y'' - 7y' + 5y = -29$$

we obtain

$$(2D^2 - 7D + 5)Dy = D(-29) = 0$$

or

$$(2D-5)(D-1)Dy = 0.$$

The roots of the auxiliary equation are 5/2, 1, and 0. Thus

$$y = \underbrace{c_1 e^{5x/2} + c_2 e^x}_{y_c} + \underbrace{c_3}_{y_p}.$$

Identifying the form of y_p as $y_p = A$ and substituting into the differential equation, we find

$$5A = -29 \qquad \text{or} \qquad A = -29/5.$$

The solution of the differential equation is

$$y = c_1 e^{5x/2} + c_2 e^x - \frac{29}{5}.$$

36. From the auxiliary equation

$$m^2 + 4 = 0$$

we find the complimentary function

$$y_c = c_1 \cos 2x + c_2 \sin 2x.$$

An annihilator for $4\cos x + 3\sin x - 8$ is $D(D^2+1)$. Applying this to both sides of the differential equation, we obtain

$$(D^2+4)D(D^2+1)y = 0.$$

The roots of the auxiliary equation are $\pm 2i$, 0, and $\pm i$. Thus

$$y = \underbrace{c_1 \cos 2x + c_2 \sin 2x}_{y_c} + \underbrace{c_3 + c_4 \cos x + c_5 \sin x}_{y_p},$$

and so a particular solution has the form

$$y_p = A + B\cos x + C\sin x.$$

Substituting this into the differential equation gives

$$4A + 3B\cos x + 3C\sin x = 4\cos x + 3\sin x - 8.$$

Equating coefficients, we find $A = -2$, $B = 4/3$, and $C = 1$. The solution of the differential equation is

$$y = c_1 \cos 2x + c_2 \sin 2x - 2 + \frac{4}{3}\cos x + \sin x.$$

46. From the auxiliary equation

$$m^2 + 4 = 0$$

we find the complimentary function

$$y_c = c_1 \cos 2x + c_2 \sin 2x.$$

Writing $\cos^2 x = (1 + \cos 2x)/2$ we see that an annihilator for $\cos^2 x$ is $D(D^2 + 4)$. Applying this to both sides of the differential equation, we have

$$D(D^2 + 4)^2 = 0.$$

The roots of the auxiliary equation are 0, $\pm 2i$, and $\pm 2i$. Thus, y has the form

$$y = \underbrace{c_1 \cos 2x + c_2 \sin 2x}_{y_c} + \underbrace{c_3 x \cos 2x + c_4 x \sin 2x + c_5}_{y_p},$$

and so a particular solution has the form

$$y_p = Ax \cos 2x + Bx \sin 2x + C.$$

Using

$$y_p'' = -4Ax \cos 2x - 4A \sin 2x - 4Bx \sin 2x + 4B \cos 2x$$

and substituting into the differential equation gives

$$-4A \sin 2x + 4B \cos 2x + 4C = \frac{1}{2} + \frac{1}{2} \cos 2x.$$

Equating coefficients, we find $A = 0$, $B = 1/8$, and $C = 1/8$. The solution of the differential equation is

$$y = c_1 \cos 2x + c_2 \sin 2x + \frac{1}{8} x \sin 2x + \frac{1}{8}.$$

54. From the auxiliary equation

$$m^4 - 5m^2 + 4 = (m^2 - 4)(m^2 - 1) = (m - 2)(m + 2)(m - 1)(m + 1) = 0$$

we find the complimentary function

$$y_c = c_1 e^{2x} + c_2 e^{-2x} + c_3 e^{-x} + c_4 e^{x}.$$

Writing $2 \cosh x = e^x + e^{-x}$ we see that an annihilator for $2 \cosh x - 6$ is $D(D-1)(D+1)$. Applying this to both sides of the differential equation, we have

$$(D - 2)(D + 2)(D - 1)^2(D + 1)^2 Dy = 0.$$

The roots of the auxiliary equation are ± 2, ± 1, ± 1, and 0. Thus, y has the form

$$y = \underbrace{c_1 e^{2x} + c_2 e^{-2x} + c_3 e^{-x} + c_4 e^{x}}_{y_c} + \underbrace{c_5 x e^{x} + c_6 x e^{-x} + c_7}_{y_p},$$

and so a particular solution has the form

$$y_p = Axe^x + Bxe^{-x} + C.$$

The necessary derivatives are

$$y_p'' = Axe^x + 2Ae^x + Bxe^{-x} - 2Be^{-x}$$
$$y_p^{(4)} = Axe^x + 4Ae^x + Bxe^{-x} - 4Be^{-x}.$$

Substituting into the differential equation gives

$$-6Ae^x + 6Be^{-x} + 4C = e^x + e^{-x} - 6.$$

Equating coefficients, we find $A = -1/6$, $B = 1/6$, and $C = -3/2$. The solution of the differential equation is

$$y = c_1 e^{2x} + c_2 e^{-2x} + c_3 e^{-x} + c_4 e^{x} - \frac{1}{6} x e^{x} + \frac{1}{6} x e^{-x} - \frac{3}{2}.$$

62. From the auxiliary equation

$$m^4 - m^3 = m^3(m-1) = 0$$

we find the complimentary function

$$y_c = c_1 + c_2 x + c_3 x^2 + c_4 e^x.$$

An annihilator for $x + e^x$ is $D^2(D-1)$. Applying this to both sides of the equation, we have

$$D^3(D-1)D^2(D-1) = 0$$

or

$$D^5(D-1)^2 = 0.$$

The roots of the auxiliary equation are 0, 0, 0, 0, 0, 1, and 1. Thus, y has the form

$$y = \underbrace{c_1 + c_2 x + c_3 x^2 + c_4 e^{x}}_{y_c} + \underbrace{c_5 x^3 + c_6 x^4 + c_7 x e^{x}}_{y_p},$$

and so a particular solution has the form

$$y_p = Ax^3 + Bx^4 + Cxe^x.$$

Using

$$y_p''' = 6A + 24Bx + Cxe^x + 3Ce^x$$

and

$$y_p^{(4)} = 24B + Cxe^x + 4Ce^x$$

and substituting into the differential equation gives

$$24B - 6A - 24Bx + Ce^x = x + e^x.$$

Equating coefficients, we find $A = -1/6$, $B = -1/24$, and $C = 1$. The general solution of the differential equation is

$$y = c_1 + c_2x + c_3x^2 + c_4e^x - \frac{1}{6}x^3 - \frac{1}{24}x^4 + xe^x.$$

From the initial conditions $y(0) = y'(0) = y''(0) = y'''(0) = 0$ and using

$$y' = c_2 + 2c_3x + c_4e^x - \frac{1}{2}x^2 - \frac{1}{6}x^3 + xe^x + e^x$$

$$y'' = 2c_3 + c_4e^x - x - \frac{1}{2}x^2 + xe^x + 2e^x$$

$$y''' = c_4e^x - 1 - x + xe^x + 3e^x$$

we have the system of equations

$$c_1 + c_4 = 0$$

$$c_2 + c_4 + 1 = 0$$

$$2c_3 + c_4 + 2 = 0$$

$$c_4 - 1 + 3 = 0.$$

The solution is $c_1 = 2$, $c_2 = 1$, $c_3 = 0$, and $c_4 = -2$. The solution of the initial value problem is

$$y = 2 + x - 2e^x - \frac{1}{6}x^3 - \frac{1}{24}x^4 + xe^x.$$

Exercises 4.5

4. Solving the auxiliary equation, we obtain

$$m^2 + 1 = 0$$
$$m_1 = -i, \quad m_2 = i$$
$$y_c = c_1 \cos x + c_2 \sin x.$$

Letting $y_1 = \cos x$ and $y_2 = \sin x$ we find the Wronskian $W(\cos x, \sin x) = 1$. Now

$$u_1' = -\frac{y_2 f}{W} = -\sin x(\sec x \tan x)$$
$$= -\tan^2 x = 1 - \sec^2 x.$$

Integrating,

$$u_1 = x - \tan x.$$

From

$$u_2' = \frac{y_1 f}{W} = \cos x(\sec x \tan x) = \tan x$$

we obtain

$$u_2 = -\ln|\cos x|.$$

Thus

$$y_p = (x - \tan x)\cos x - \ln|\cos x|(\sin x)$$
$$= x\cos x - \sin x - \sin x \ln|\cos x|$$

and

$$y = y_c + y_p = c_1 \cos x + c_2 \sin x + x\cos x - \sin x - \sin x \ln|\cos x|.$$

Combining $c_2 \sin x$ and $\sin x$ we can express the solution as

$$y = c_1 \cos x + c_3 \sin x + x \cos x - \sin x \ln|\cos x|.$$

14. Solving the auxiliary equation, we obtain

$$m^2 - 2m + 1 = 0$$
$$(m-1)^2 = 0$$
$$m = 1,\ 1$$

and

$$y_c = c_1 e^x + c_2 x e^x.$$

Letting $y_1 = e^x$ and $y_2 = xe^x$ we find the Wronskian $W(e^x, xe^x) = e^{2x}$. Now

$$u_1' = -\frac{xe^x e^x \tan^{-1} x}{e^{2x}} = -x \tan^{-1} x,$$

so

$$u_1 = -\frac{1+x^2}{2} \tan^{-1} x + \frac{x}{2}.$$

From

$$u_2' = \frac{e^x e^x \tan^{-1} x}{e^{2x}} = \tan^{-1} x$$

we obtain

$$u_2 = x \tan^{-1} x - \frac{1}{2} \ln(1+x^2).$$

Thus

$$\begin{aligned} y_p &= \left(-\frac{1+x^2}{2} \tan^{-1} x + \frac{x}{2}\right) e^x + \left(x \tan^{-1} x - \frac{1}{2} \ln(1+x^2)\right) x e^x \\ &= \frac{xe^x}{2} - \frac{e^x}{2}(1+x^2) \tan^{-1} x + x^2 e^x \tan^{-1} x - \frac{e^x}{2} \ln(1+x^2) \\ &= \frac{xe^x}{2} - \frac{e^x}{2} \tan^{-1} x + \frac{x^2 e^x}{2} \tan^{-1} x - \frac{e^x}{2} \ln(1+x^2) \end{aligned}$$

and

$$\begin{aligned} y &= y_c + y_p \\ &= c_1 e^x + c_2 x e^x + \frac{1}{2} x e^x - \frac{1}{2} e^x [\tan^{-1} x - x^2 \tan^{-1} x + \ln(1+x^2)] \\ &= c_1 e^x + c_3 x e^x - \frac{1}{2} e^x [(1-x^2) \tan^{-1} x + \ln(1+x^2)]. \end{aligned}$$

22. Solving the auxiliary equation, we obtain

$$m^3 + 4m = m(m^2+4) = 0$$

$$m = 0,\ -2i,\ 2i$$

and

$$y_c = c_1 + c_2 \cos 2x + c_3 \sin 2x.$$

Letting $y_1 = 1$, $y_2 = \cos 2x$, and $y_3 = \sin 2x$ we find the Wronskian

$$W(1, \cos 2x, \sin 2x) = \begin{vmatrix} 1 & \cos 2x & \sin 2x \\ 0 & -2\sin 2x & 2\cos 2x \\ 0 & -4\cos 2x & -4\sin 2x \end{vmatrix} = 8.$$

Now

$$u_1' = \frac{W_1}{W} = \frac{1}{8}\begin{vmatrix} 0 & \cos 2x & \sin 2x \\ 0 & -2\sin 2x & 2\cos 2x \\ \sec 2x & -4\cos 2x & -4\sin 2x \end{vmatrix}$$

$$= \frac{1}{8}(2\sec 2x) = \frac{1}{4}\sec 2x$$

so

$$u_1 = \frac{1}{8}\ln|\sec 2x + \tan 2x|;$$

$$u_2' = \frac{W_2}{W} = \frac{1}{8}\begin{vmatrix} 1 & 0 & \sin 2x \\ 0 & 0 & 2\cos 2x \\ 0 & \sec 2x & -4\sin 2x \end{vmatrix} = \frac{1}{8}(-2) = -\frac{1}{4}$$

so

$$u_2 = -\frac{1}{4}x;$$

and

$$u_3' = \frac{W_3}{W} = \frac{1}{8}\begin{vmatrix} 1 & \cos 2x & 0 \\ 0 & -2\sin 2x & 0 \\ 0 & -4\cos 2x & \sec 2x \end{vmatrix}$$

$$= \frac{1}{8}(-2\sin 2x\sec 2x) = -\frac{1}{4}\tan 2x$$

so

$$u_3 = \frac{1}{8}\ln|\cos 2x|.$$

Thus

$$y_p = \frac{1}{8}\ln|\sec 2x + \tan 2x| - \frac{1}{4}x\cos 2x + \frac{1}{8}\ln|\cos 2x|(\sin 2x)$$

and

$$\begin{aligned} y &= y_c + y_p \\ &= c_1 + c_2\cos 2x + c_3\sin 2x + \frac{1}{8}\ln|\sec 2x + \tan 2x| \\ &\quad - \frac{1}{4}x\cos 2x + \frac{1}{8}\sin 2x\ln|\cos 2x|. \end{aligned}$$

Since the tangent function is undefined at $\pm\pi/2$ we restrict x to the interval $(-\pi/4, \pi/4)$.

26. Solving the auxiliary equation, we obtain

$$2m^2 + m - 1 = 0$$

$$(2m-1)(m+1) = 0$$

$$m_1 = \frac{1}{2}, \quad m_2 = -1$$

and

$$y_c = c_1 e^{x/2} + c_2 e^{-x}.$$

Letting $y_1 = e^{x/2}$ and $y_2 = e^{-x}$ we find the Wronskian $W(e^{x/2}, e^{-x}) = -\frac{3}{2}e^{-x/2}$. Now

$$u_1' = -\frac{e^{-x}\frac{1}{2}(x+1)}{-\frac{3}{2}e^{-x/2}} = \frac{1}{3}e^{-x/2}(x+1)$$

so

$$u_1 = e^{x/2}\left(-\frac{2}{3}x - 2\right),$$

and

$$u_2' = \frac{e^{x/2}\frac{1}{2}(x+1)}{-\frac{3}{2}e^{-x/2}} = -\frac{1}{3}e^{x}(x+1)$$

so

$$u_2 = -\frac{1}{3}xe^{x}.$$

Thus

$$y_p = e^{-x/2}\left(-\frac{2}{3}x - 2\right)e^{x/2} + \left(-\frac{1}{3}xe^{x}\right)e^{-x} = -x - 2$$

and

$$y = y_c + y_p = c_1 e^{x/2} + c_2 e^{-x} - x - 2.$$

Applying the initial conditions $y(0) = 1$, $y'(0) = 0$ to this general solution gives $c_1 = 8/3$ and $c_2 = 1/3$. Hence the solution to the initial value problem is

$$y = \frac{8}{3}e^{x/2} + \frac{1}{3}e^{-x} - x - 2.$$

30. Write the equation in the form

$$y'' - \frac{4}{x}y' + \frac{6}{x^2}y = \frac{1}{x^3}$$

and identify $f(x) = 1/x^3$. From $y_1 = x^2$ and $y_2 = x^3$ we compute

$$W(y_1, y_2) = \begin{vmatrix} x^2 & x^3 \\ 2x & 3x^2 \end{vmatrix} = 3x^4 - 2x^4 = x^4.$$

Now

$$u_1' = -\frac{x^3/x^3}{x^4} = -\frac{1}{x^4}$$

so

$$u_1 = \frac{1}{3x^3},$$

and

$$u_2' = \frac{x^2/x^3}{x^4} = \frac{1}{x^5}$$

so

$$u_2 = \frac{1}{4x^4}.$$

Thus,

$$y_p = \frac{x^2}{3x^3} - \frac{x^3}{4x^4} = \frac{1}{12x}$$

and

$$y = y_c + y_p = c_1x^2 + c_2x^3 + \frac{1}{12x}.$$

5 Applications of Second-Order Differential Equations: Vibrational Models

Exercises 5.1

4. Write the differential equation in the form

$$x'' + 16x = 0.$$

Then

$$x(t) = c_1 \cos 4t + c_2 \sin 4t$$

and

$$x'(t) = -4c_1 \sin 4t + 4c_2 \cos 4t.$$

Using the initial conditions, we find

$$1 = x(0) = c_1$$

and

$$-2 = x'(0) = 4c_2.$$

Thus, $c_1 = 1$, $c_2 = -1/2$, and

$$x(t) = \cos 4t - \frac{1}{2}\sin 4t.$$

Now

$$A = \sqrt{c_1^2 + c_2^2} = \sqrt{1 + \frac{1}{4}} = \frac{\sqrt{5}}{2}$$

and

$$\tan\phi = \frac{c_1}{c_2} = -2.$$

The inverse tangent of -2 is approximately -1.107, which is a fourth quadrant angle. Since $\sin\phi = c_1/A = 2/\sqrt{5} > 0$ and $\cos\phi = c_2/A = -1/\sqrt{5} < 0$, we see that ϕ is in the second quadrant. Thus

$$\phi = \tan^{-1}(-2) + \pi \approx -1.107 + \pi \approx 2.034$$

and

$$x(t) = A\sin(4t + \phi) \approx \frac{\sqrt{5}}{2}\sin(4t + 2.034).$$

12. We know that frequency is given by $f = \omega/2\pi$. Thus, $\omega = 2\pi f$. Since we are given $f = 2/\pi$, we have $\omega = 4$. Now, we also know that $\omega^2 = k/m$ or $k = m\omega^2$. Since $m = 20$ we see that $k = 320$. If, instead, $m = 80$, then

$$f = \frac{\omega}{2\pi} = \frac{\sqrt{k/m}}{2\pi} = \frac{\sqrt{320/80}}{2\pi} = \frac{1}{\pi}.$$

18. From $F = ks$, we have $32 = k(2)$ and $k = 16$. Since $m = W/g = 32/32 = 1$ and $w^2 = k/m = 16$, the differential equation of motion is

$$x'' + 16x = 0.$$

Hence

$$x(t) = c_1 \cos 4t + c_2 \sin 4t.$$

The initial conditions $x(0) = -1$ and $x'(0) = -2$ give $c_1 = -1$ and $c_2 = -\frac{1}{2}$, respectively. Thus,

$$x(t) = -\cos 4t - \frac{1}{2}\sin 4t,$$

and the amplitude of motion is

$$A = \sqrt{(-1)^2 + \left(-\frac{1}{2}\right)^2} = \frac{\sqrt{5}}{2}.$$

Since the period of oscillations is $2\pi/4 = \pi/2$ the weight will have completed 8 complete cycles in 4π seconds.

22. Using $m = 1$ and $k = 9$ we see that $\omega = \sqrt{k/m} = 3$. The equation of motion then has the form

$$x(t) = c_1 \cos 3t + c_2 \sin 3t.$$

The initial conditions $x(0) = -1$ and $x'(0) = -\sqrt{3}$ give $c_1 = -1$ and $c_2 = -\sqrt{3}/3$. Hence,

$$x(t) = -\cos 3t - \frac{\sqrt{3}}{3}\sin 3t.$$

Now

$$A = \sqrt{c_1^2 + c_2^2} = \sqrt{1 + \frac{1}{3}} = \frac{2}{\sqrt{3}}$$

and

$$\tan\phi = \frac{c_1}{c_2} = \frac{-1}{-\sqrt{3}/3} = \sqrt{3}.$$

Thus, $\phi = \pi/3$ or $\phi = 4\pi/3$. Since $c_1 < 0$ and $c_2 < 0$, $\sin\phi$ and $\cos\phi$ are negative and ϕ is in the third quadrant. Thus, $\phi = 4\pi/3$ and

$$x(t) = A\sin(\omega t + \phi) = \frac{2}{\sqrt{3}}\sin\left(3t + \frac{4\pi}{3}\right).$$

Now

$$x'(t) = 2\sqrt{3}\cos\left(3t + \frac{4\pi}{3}\right),$$

and we want to find t so that $x'(t) = 3$. From

$$3 = 2\sqrt{3}\cos\left(3t + \frac{4\pi}{3}\right)$$

we find

$$\frac{\sqrt{3}}{2} = \cos\left(3t + \frac{4\pi}{3}\right)$$

and

$$3t + \frac{4\pi}{3} = \frac{\pi}{6} + 2k\pi \quad \text{or} \quad 3t + \frac{4\pi}{3} = -\frac{\pi}{6} + 2k\pi.$$

Solving for t we have

$$t = -\frac{7\pi}{18} + \frac{2k\pi}{3}, \quad k = 1,\ 2,\ 3,\ \ldots$$

$$t = -\frac{\pi}{2} + \frac{2k\pi}{3}, \quad k = 1,\ 2,\ 3,\ \ldots.$$

24. Let m denote the mass in slugs of the first weight. Let k_1 and k_2 be the spring constants and $k = 4k_1k_2/(k_1 + k_2)$ the effective spring constant of the system. Now, the numerical value of the first weight is $W = mg = 32m$, so

$$32m = k_1\left(\frac{1}{3}\right) \quad \text{and} \quad 32m = k_2\left(\frac{1}{2}\right).$$

From these equations we find $2k_1 = 3k_2$. The given period of the combined system is $2\pi/w = \pi/15$, so $w = 30$. Since the mas of an 8 pound weight is $1/4$ slug, we have from $w^2 = k/m$

$$30^2 = \frac{k}{1/4} = 4k \quad \text{or} \quad k = 225.$$

We now have the system of equations

$$\frac{4k_1k_2}{k_1 + k_2} = 225$$

$$2k_1 = 3k_2.$$

Solving the second equation for k_1 and substituting in the first equation, we obtain

$$\frac{4(3k_2/2)k_2}{3k_2/2+k_2} = \frac{12k_2^2}{5k_2} = \frac{12k_2}{5} = 225.$$

Thus, $k_2 = 375/4$ and $k_1 = 1125/8$. Finally, the value of the first weight is

$$W = 32m = \frac{k_1}{3} = \frac{1125/8}{3} = \frac{375}{8} \approx 46.88 \text{ lbs.}$$

Exercises 5.2

4. After the weight is attached, the elongation of the spring is 4 ft. Thus, $8 = k(4)$ which implies $k = 2$ lb/ft. Also, $m = 8/32 = 1/4$ slug. The differential equation is then

$$\frac{1}{4}\frac{d^2x}{dt^2} = -2x - \sqrt{2}\frac{dx}{dt}$$

or

$$\frac{d^2x}{dt^2} + 4\sqrt{2}\frac{dx}{dt} + 8x = 0.$$

Solving this equation gives

$$x(t) = c_1e^{-2\sqrt{2}\,t} + c_2e^{-2\sqrt{2}\,t}.$$

Using the initial conditions $x(0) = 0$ and $x'(0) = 5$, we find $c_1 = 0$ and $c_2 = 5$. Hence,

$$x(t) = 5te^{-2\sqrt{2}\,t}.$$

Now

$$x'(t) = -10\sqrt{2}\,te^{-2\sqrt{2}\,t} + 5e^{-2\sqrt{2}\,t}$$

$$= 5e^{-2\sqrt{2}\,t}(1 - 2\sqrt{2}\,t).$$

At the maximum displacement $x'(t) = 0$. The last equation implies $t = \sqrt{2}/4$, and so the maximum displacement is

$$x\left(\frac{\sqrt{2}}{4}\right) = \frac{5\sqrt{2}}{4}e^{-1} \approx 0.65 \text{ ft.}$$

10. From $F = ks$ and $m = W/g$ we have $24 = k(4)$ or $k = 6$ and $m = 24/32 = 3/4$ slug. The differential equation is then

$$\frac{3}{4}\frac{d^2x}{dt^2} = -6x - \beta\frac{dx}{dt}$$

or

$$\frac{d^2x}{dt^2} + \frac{4}{3}\beta\frac{dx}{dt} + 8x = 0.$$

When $\beta > 3\sqrt{2}$ the real roots of the auxiliary equation

$$m^2 + \frac{4}{3}\beta m + 8 = 0$$

are

$$m_1 = -\frac{2\beta}{3} + \frac{2}{3}\sqrt{\beta^2 - 18}\,, \quad m_2 = -\frac{2\beta}{3} - \frac{2}{3}\sqrt{\beta^2 - 18}\,.$$

Therefore,

$$\begin{aligned} x(t) &= c_1 e^{\left(-2+2\sqrt{\beta^2-18}\right)\beta t/3} + c_2 e^{\left(-2-2\sqrt{\beta^2-18}\right)\beta t/3} \\ &= e^{-2\beta t/3}\left[c_1 e^{2\sqrt{\beta^2-18}\,t/3} + c_2 e^{-2\sqrt{\beta^2-18}\,t/3}\right]. \end{aligned}$$

By redefining the arbitrary constants, the quantity in the brackets can be written as

$$x(t) = e^{-2\beta t/3}\left[C_1 \cosh \frac{2}{3}\sqrt{\beta^2 - 18}\,t + C_2 \sinh \frac{2}{3}\sqrt{\beta^2 - 18}\,t\right].$$

Now the initial conditions $x(0) = 0$ and $x'(0) = -2$ give $C_1 = 0$ and $C_2 = -3/\sqrt{\beta^2 - 18}$, respectively. Thus, we obtain

$$x(t) = \frac{-3}{\sqrt{\beta^2 - 18}}\, e^{-2\beta t/3} \sinh \frac{2}{3}\sqrt{\beta^2 - 18}\,t.$$

14. The form of the differential equation is

$$x'' + 2\lambda x' + \omega^2 x = 0$$

where $2\lambda = \beta/m$ and $\omega^2 = k/m$. We are given $m = 1$ and $k = 25$, so $2\lambda = \beta$ and $\omega^2 = 25$. The given quasi period is $\pi/2$ so

$$\frac{\pi}{2} = \frac{2\pi}{\sqrt{\omega^2 - \lambda^2}}$$

or

$$\omega^2 - \lambda^2 = 16.$$

Thus

$$25 - \lambda^2 = 16$$

and $\lambda = 3$. Therefore, the damping constant is $\beta = 2\lambda = 6$.

Exercises 5.3

4. Since the mass is 1 slug, the weight is 32 pounds. From Hooke's law $32 = k(2)$ and the spring constant is 16. We are given that the damping constant is $\beta = 8$, and the external force is $f(t) = e^{-t}\sin 4t$. The differential equation then is

$$x'' + 8x' + 16x = e^{-t}\sin 4t.$$

The auxiliary equation has a double root of -4 so the complementary function is

$$x_c(t) = c_1 e^{-4t} + c_2 t e^{-4t}.$$

Using the method of undetermined coefficients, we assume

$$x_p(t) = Ae^{-t}\cos 4t + Be^{-t}\sin 4t.$$

Then

$$x_p'(t) = (4B - A)e^{-t}\cos 4t - (4A + B)e^{-t}\sin 4t$$

and

$$x_p''(t) = -(15A + 8B)e^{-t}\cos 4t - (15B - 8A)e^{-t}\sin 4t.$$

Substituting into the differential equation, we obtain

$$(-7A + 24B)e^{-t}\cos 4t - (24A + 7B)e^{-t}\sin 4t = e^{-t}\sin 4t.$$

Thus

$$-7A + 24B = 0$$

$$-24A - 7B = 1$$

so $A = -24/625$ and $B = -7/625$. It follows that

$$x_p(t) = -\frac{24}{625}e^{-t}\cos 4t - \frac{7}{625}e^{-t}\sin 4t$$

and

$$x(t) = c_1 e^{-4t} + c_2 t e^{-4t} - \frac{1}{625}e^{-t}(24\cos 4t + 7\sin 4t).$$

Now, $x(0) = x'(0) = 0$ so $c_1 = 24/625$ and $c_2 = 100/625$. The equation of motion of the system is

$$x(t) = \frac{1}{625}e^{-4t}(24 + 100t) - \frac{1}{625}e^{-t}(24\cos 4t + 7\sin 4t).$$

As $t \to \infty$ we see that $x(t) \to 0$. Thus, for large time, the displacements approach 0.

8. We must solve

$$\frac{1}{2}x'' = -4(x - 5\cos t) - 2x'$$

or

$$x'' + 4x' + 8x = 40\cos t$$

subject to $x(0) = x'(0) = 0$. Solving the related homogeneous equation, we obtain

$$x_c(t) = e^{-2t}(c_1 \cos 2t + c_2 \sin 2t).$$

Assuming a particular solution of the form

$$x_p(t) = A\cos t + B\sin t$$

we find that A and B must satisfy

$$7A + 4B = 40$$

$$-4A + 7B = 0.$$

Solving for A and B yields $A = 56/13$ and $B = 32/13$. Thus,

$$\begin{aligned} x &= x_c + x_p \\ &= e^{-2t}(c_1 \cos 2t + c_2 \sin 2t) + \frac{56}{13}\cos t + \frac{32}{13}\sin t. \end{aligned}$$

The initial conditions then give $c_1 = -56/13$ and $c_2 = -72/13$. Therefore,

$$x(t) = e^{-2t}\left(-\frac{56}{13}\cos 2t - \frac{72}{13}\sin 2t\right) + \frac{56}{13}\cos t + \frac{32}{13}\sin t.$$

12. (a) Recall $\lambda = \beta/2m$ and $\omega^2 = k/m$. In the case when $k = 3$ and $m = 1$ then $\lambda = \beta/2$ and $\omega^2 = 3$. The underdamped system is characterized by $\omega^2 - \lambda^2 > 0$ or $3 - \beta^2/4 > 0$. This latter inequality implies $0 < \beta < \sqrt{12}$ or $0 < \beta < 2\sqrt{3}$. Now, the system is in resonance when $\gamma = \sqrt{\omega^2 - 2\lambda^2}$. In order to have $\omega^2 - 2\lambda^2 > 0$ or $3 - \beta^2/2 > 0$ we must further require that $0 < \beta < \sqrt{6}$.

    (b) When $F_0 = 3$, the resonance curve is given by

$$g(\gamma) = \frac{3}{\sqrt{(3-\gamma^2)^2 + \beta^2\gamma^2}},$$

and the family of graphs is shown for various values of β.

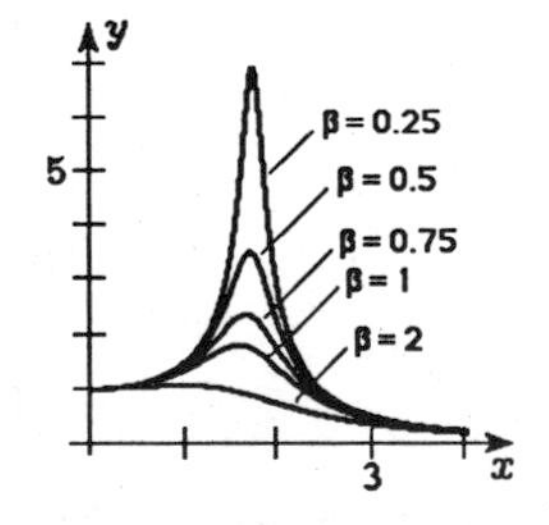

Exercises 5.4

4. Since
$$R^2 - 4L/C = (20)^2 - 4(1)/0.01 = 400 - 400 = 0,$$
the circuit is critically damped.

8. The differential equation is
$$\frac{d^2q}{dt^2} + 100\frac{dq}{dt} + 2500q = 30.$$
The auxiliary equation is
$$m^2 + 100m + 2500 = (m+50)^2 = 0,$$
so
$$q_c(t) = c_1 e^{-50t} + c_2 te^{-50t}.$$
To find $q_p(t)$ we assume a particular solution of the form $q_p(t) = A$. Then $2500A = 30$ and $A = 3/250 = 0.012$. Thus,
$$q(t) = q_c + q_p = c_1 e^{-50t} + c_2 te^{-50t} + 0.012$$
and
$$i(t) = \frac{dq}{dt} = -50c_1 e^{-50t} - 50c_2 te^{-50t} + c_2 e^{-50t}.$$
Letting $q(0) = 0$ and $i(0) = 2$ we obtain
$$c_1 + 0.012 = 0$$
$$-50c_1 + c_2 = 2.$$
This gives $c_1 = -0.012$ and $c_2 = 1.4$. The charge on the capacitor is
$$q(t) = -0.012e^{-50t} + 1.4te^{-50t} + 0.012,$$
and the current in the circuit is
$$\begin{aligned} i(t) &= 0.6e^{-50t} - 70te^{-50t} + 1.4e^{-50t} \\ &= (2 - 70t)e^{-50t}. \end{aligned}$$
To find the maximum charge we solve $i(t) = 0$, obtaining $t = 1/35$. We observe also that for $t < 1/35$, $i(t) > 0$ and for $t > 1/35$, $i(t) < 0$, so that $q(t)$ is maximum at $t = 1/35$. Finally, the maximum charge is
$$q(1/35) = -0.012e^{-50/35} + 1.4(1/35)e^{-50/35} + 0.012 \approx 0.01871 \text{ coulombs}.$$

12. The differential equation is

$$\frac{1}{2}\frac{d^2q}{dt^2} + 20\frac{dq}{dt} + 1000q = 100\sin 60t + 200\cos 40t$$

or

$$\frac{dq^2}{dt^2} + 40\frac{dq}{dt} + 2000q = 200\sin 60 + 400\cos 40t.$$

We assume a particular solution of the form

$$q_p = A\sin 60t + B\cos 60t + C\sin 40t + D\cos 40t.$$

Then

$$q_p' = 60A\cos 60t - 60B\sin 60t + 40C\cos 40t - 40D\sin 40t$$

and

$$q_p'' = -3600A\sin 60t - 3600B\cos 60t - 1600C\sin 40t - 1600D\cos 40t.$$

Substituting into the differential equation, we obtain

$$\begin{aligned}
-3600A\sin 60t &- 3600B\cos 60t - 1600C\sin 40t - 1600D\cos 40t \\
&+ 2400A\cos 60t - 2400B\sin 60t + 1600C\cos 40t - 1600D\sin 40t \\
&+ 2000A\sin 60t + 2000B\cos 60t + 2000C\sin 40t + 2000D\cos 40t \\
&= (-1600A - 2400B)\sin 60t + (2400A - 1600B)\cos 60t \\
&\quad + (400C - 1600D)\sin 40t + (1600C + 400D)\cos 40t \\
&= 200\sin 60t + 400\cos 40t.
\end{aligned}$$

Equating corresponding coefficients, we obtain

$$\begin{aligned}-1600A - 2400B &= 200\\ 2400A - 1600B &= 0\end{aligned} \quad\text{and}\quad \begin{aligned}400C - 1600D &= 0\\ 1600C + 400D &= 400\end{aligned}$$

or

$$\begin{aligned}8A + 12B &= -1\\ 3A - 2B &= 0\end{aligned} \quad\text{and}\quad \begin{aligned}C - 4D &= 0\\ 4C + D &= 1.\end{aligned}$$

Solving, we obtain $A = -1/26$, $B = -3/52$, $C = 4/17$ and $D = 1/17$. The steady-state solution is

$$q_p = -\frac{1}{26}\sin 60t - \frac{3}{52}\cos 60t + \frac{4}{17}\sin 40t + \frac{1}{17}\cos 40t$$

and the steady-state current is

$$i_p(t) = \frac{dq_p}{dt} = -\frac{30}{13}\cos 60t + \frac{45}{13}\sin 60t + \frac{160}{17}\cos 40t - \frac{40}{17}\sin 40t.$$

6 Differential Equations with Variable Coefficients

Exercises 6.1

8. Substitute $y = x^m$ into the differential equation $x^2y'' + 3xy' - 4y = 0$ and obtain

$$\begin{aligned} x^2 \cdot m(m-1)x^{m-2} + 3x \cdot mx^{m-1} - 4x^m &= m(m-1)x^m + 3mx^m - 4x^m \\ &= x^m[m(m-1) + 3m - 4] \\ &= x^m[m^2 + 2m - 4] \\ &= 0. \end{aligned}$$

Now set $m^2 + 2m - 4 = 0$ and solve by the quadratic formula:

$$m = \frac{-2 \pm \sqrt{4+16}}{2} = -1 \pm \sqrt{5}.$$

Thus $m_1 = -1 - \sqrt{5}$ and $m_2 = -1 + \sqrt{5}$ so that

$$y = c_1x^{-1-\sqrt{5}} + c_2x^{-1+\sqrt{5}}.$$

18. Substitute $y = x^m$ into the differential equation $x^3y''' + xy' - y = 0$ and obtain

$$\begin{aligned} x^3 \cdot m(m-1)(m-2)x^{m-3} + x \cdot mx^{m-1} - x^m &= m(m-1)(m-2)x^m + mx^m - x^m \\ &= x^m[m^3 - 3m^2 + 2m + m - 1] \\ &= x^m[m^3 - 3m^2 + 3m - 1] \\ &= x^m(m-1)^3 \\ &= 0. \end{aligned}$$

Thus, $m_1 = m_2 = m_3 = 1$ is a root of multiplicity 3. It follows that three linearly independent solutions are $y_1 = x$, $y_2 = x \ln x$, and $y_3 = x(\ln x)^2$. Therefore, the general solution of the equation is

$$y = c_1x + c_2x \ln x + c_3x(\ln x)^2.$$

24. The substitution $y = x^m$ leads to the auxiliary equation

$$m(m-1) - 5m + 8 = m^2 - 6m + 8 = (m-2)(m-4) = 0.$$

Thus, $m_1 = 2$ and $m_2 = 4$ so that

$$y = c_1 x^2 + c_2 x^4$$

and

$$y' = 2c_1 x + 4c_2 x^3.$$

From $y(2) = 32$ and $y'(2) = 0$ we obtain

$$\begin{aligned} 4c_1 + 16c_2 &= 32 \\ 4c_1 + 32c_2 &= 0 \end{aligned} \qquad \text{or} \qquad \begin{aligned} c_1 + 4c_2 &= 8 \\ c_1 + 8c_2 &= 0. \end{aligned}$$

Solving this system, we find $c_1 = 16$ and $c_2 = -2$ so that

$$y = 16x^2 - 2x^4.$$

28. We use the substitution $t = -x$ since the initial conditions are on the interval $(-\infty, 0)$. Then

$$\frac{dy}{dt} = \frac{dy}{dx}\frac{dx}{dt} = -\frac{dy}{dx}$$

and

$$\begin{aligned} \frac{d^2y}{dt^2} &= \frac{d}{dt}\left(\frac{dy}{dt}\right) = \frac{d}{dt}\left(-\frac{dy}{dx}\right) = -\frac{d}{dt}(y') \\ &= -\frac{dy'}{dx}\frac{dx}{dt} = -\frac{d^2y}{dx^2}\frac{dx}{dt} = \frac{d^2y}{dx^2}. \end{aligned}$$

The differential equation and initial conditions become, respectively

$$t^2\frac{d^2y}{dt^2} - 4t\frac{dy}{dt} + 6y = 0$$

$$y(2) = 8, \quad y'(2) = 0.$$

The substitution $y = t^m$ leads to the auxiliary equation

$$m(m-1) - 4m + 6 = m^2 - 5m + 6 = (m-2)(m-3) = 0.$$

Thus $m_1 = 2$ and $m_2 = 3$ so that

$$y = c_1 t^2 + c_2 t^3.$$

The initial conditions then yield

$$\begin{aligned} 4c_1 + 8c_2 &= 8 \\ 4c_1 + 12c_2 &= 0 \end{aligned} \qquad \text{or} \qquad \begin{aligned} c_1 + 2c_2 &= 2 \\ c_1 + 3c_2 &= 0 \end{aligned}$$

from which we find $c_1 = 6$ and $c_2 = -2$. Hence

$$\begin{aligned} y &= 6t^2 - 2t^3 \\ &= 6(-x)^2 - 2(-x)^3 \\ &= 6x^2 + 2x^3, \quad x < 0. \end{aligned}$$

32. The substitution $y = x^m$ leads to the auxiliary equation

$$m^2 - 3m + 2 = (m-1)(m-2) = 0.$$

Then $m_1 = 1$ and $m_2 = 2$ so that

$$y_c = c_1 x + c_2 x^2.$$

Letting $y_1 = x$ and $y_2 = x^2$ we obtain

$$W = \begin{vmatrix} x & x^2 \\ 1 & 2x \end{vmatrix} = x^2.$$

To use variation of parameters we write the differential equation in the form

$$y'' - \frac{2}{x}y' + \frac{2}{x^2}y = x^2 e^x$$

and identify $f(x) = x^2 e^x$. Then

$$\begin{aligned} u_1 &= -\int \frac{y_2 f}{W}\,dx \\ &= -\int \frac{x^2 x^2 e^x}{x^2}\,dx \\ &= -\int x^2 e^x\,dx \\ &= -x^2 e^x + 2xe^x - 2e^x, \end{aligned}$$

and

$$\begin{aligned} u_2 &= \int \frac{y_1 f}{W}\,dx \\ &= \int \frac{x x^2 e^x}{x^2}\,dx \\ &= \int xe^x\,dx \\ &= xe^x - e^x, \end{aligned}$$

so

$$y_p = -x^3e^x + 2x^2e^x - 2xe^x + x^3e^x - x^2e^x$$
$$= x^2e^x - 2xe^x.$$

The general solution is

$$y = c_1x + c_2x^2 + x^2e^x - 2xe^x.$$

38. From Example 6 in the text we see that

$$\frac{dy}{dx} = \frac{1}{x}\frac{dy}{dt} \quad \text{and} \quad \frac{d^2y}{dx^2} = \frac{1}{x^2}\left[\frac{d^2y}{dt^2} - \frac{dy}{dt}\right].$$

Substituting into the differential equation, we obtain

$$2\frac{d^2y}{dx^2} - 2\frac{dy}{dt} - 3\frac{dy}{dt} - 3y = 1 + 2e^t + e^{2t}$$

or

$$2\frac{d^2y}{dx^2} - 5\frac{dy}{dt} - 3y = 1 + 2e^t + e^{2t}.$$

This is an equation with constant coefficients. Its auxiliary equation is

$$2m^2 - 5m - 3 = (2m + 1)(m - 3) = 0.$$

Thus $m_1 = -1/2$ and $m_2 = 3$ so that

$$y_c = c_1e^{-t/2} + c_2e^{3t}.$$

Using undetermined coefficients we try

$$y_p = A + Be^t + Cc^{2t}.$$

Then

$$y_p' = Be^t + 2Ce^{2t}$$
$$y_p'' = Be^t + 4Ce^{2t}.$$

Substituting,

$$2Be^t + 8Ce^{2t} - 5Be^t - 10Ce^{2t} - 3A - 3Be^t - 3Ce^{2t} = 1 + 2e^t + e^{2t}$$

or

$$-3A - 6Be^t - 5Ce^{2t} = 1 + 2e^t + e^{2t}.$$

Thus, $A = -1/3$, $B = -1/3$, and $C = -1/5$, so that

$$y_p = -\frac{1}{3} - \frac{1}{3}e^t - \frac{1}{5}e^{2t}.$$

Hence

$$y = y_c + y_p = c_1 e^{-t/2} + c_2 e^{3t} - \frac{1}{3} - \frac{1}{3}e^t - \frac{1}{5}e^{2t}$$

and the general solution of the original differential equation is

$$y = c_1 x^{-1/2} + c_2 x^3 - \frac{1}{3} - \frac{1}{3}x - \frac{1}{5}x^2.$$

42. The differential equation

$$(3x+4)^2 y'' + 10(3x+4)y' + 9y = 0$$

is not a Cauchy-Euler equation. Make the substitution $t = 3x + 4$. Then

$$y' = \frac{dy}{dx} = \frac{dy}{dt}\frac{dt}{dx} = \frac{dy}{dt} \cdot 3 = 3\frac{dy}{dt}$$

and

$$y'' = \frac{dy'}{dx} = \frac{dy'}{dt}\frac{dt}{dx} = \frac{dy'}{dt} \cdot 3 = 3\frac{d}{dt}(y') = 3\frac{d}{dt}\left(3\frac{dy}{dt}\right) = 9\frac{d^2y}{dt^2}.$$

Now write the differential equation as

$$t^2 \cdot 9\frac{d^2y}{dt^2} + 10t \cdot 3\frac{dy}{dt} + 9y = 0$$

or

$$9t^2\frac{d^2y}{dt^2} + 30t\frac{dy}{dt} + 9y = 0.$$

This is a Cauchy-Euler equation with auxiliary equation

$$9m^2 + 21m + 9 = 3m^2 + 7m + 3 = 0.$$

From the quadratic formula we have

$$m = \frac{-7 \pm \sqrt{49-36}}{6} = \frac{-7 \pm \sqrt{13}}{6}.$$

Thus $m_1 = (-7 - \sqrt{13})/6$ and $m_2 = (-7 + \sqrt{13})/6$, so

$$y = c_1 t^{-(7-\sqrt{13})/6} + c_2 t^{(-7+\sqrt{13})/6}$$

or

$$y = c_1 (3x+4)^{-(7-\sqrt{13})/6} + c_2 (3x+4)^{(-7+\sqrt{13})/6}.$$

Exercises 6.2

8. The auxiliary equation is $m^2 - 1 = 0$, so the solution is

$$y = c_1 e^x + c_2 e^{-x}.$$

To solve with power series we assume a solution of the form

$$y = \sum_{n=0}^{\infty} c_n x^n.$$

Then

$$y'' = \sum_{n=2}^{\infty} n(n-1)c_n x^{n-2}$$

and

$$y'' - y = \underbrace{\sum_{n=2}^{\infty} n(n-1)c_n x^{n-2}}_{k=n-2} - \underbrace{\sum_{n=0}^{\infty} c_n x^n}_{k=n}$$

$$= \sum_{k=0}^{\infty}(k+2)(k+1)c_{k+2}x^k - \sum_{k=0}^{\infty} c_k x^k$$

$$= \sum_{k=0}^{\infty}[(k+2)(k+1)c_{k+2} - c_k]x^k = 0.$$

Thus

$$(k+2)(k+1)c_{k+2} - c_k = 0, \qquad k = 0, 1, 2, \ldots$$

and the recurrence relation is

$$c_{k+2} = \frac{c_k}{(k+2)(k+1)}, \qquad k = 0, 1, 2, \ldots.$$

Iterating, we find

$$c_2 = \frac{c_0}{2 \cdot 1} = \frac{1}{2}c_0$$

$$c_3 = \frac{c_1}{3 \cdot 2} = \frac{1}{3 \cdot 2}c_1$$

$$c_4 = \frac{c_2}{4 \cdot 3} = \frac{1}{4 \cdot 3 \cdot 2}c_0$$

$$c_5 = \frac{c_3}{5 \cdot 4} = \frac{1}{5 \cdot 4 \cdot 3 \cdot 2}c_1$$

$$c_6 = \frac{c_4}{6 \cdot 5} = \frac{1}{6 \cdot 5 \cdot 4 \cdot 3 \cdot 2}c_0$$

$$c_7 = \frac{c_5}{7 \cdot 6} = \frac{1}{7 \cdot 6 \cdot 5 \cdot 4 \cdot 3 \cdot 2}c_1.$$

Therefore,

$$\begin{aligned}
y &= c_0 + c_1 x + c_2 x^2 + c_3 x^3 + \dots \\
&= c_0 + c_1 x + \frac{1}{2} c_0 x^2 + \frac{1}{3 \cdot 2} c_1 x^3 + \frac{1}{4 \cdot 3 \cdot 2} c_0 x^4 + \frac{1}{5 \cdot 4 \cdot 3 \cdot 2} c_1 x^5 + \dots \\
&= c_0 \left[1 + \frac{1}{2!} x^2 + \frac{1}{4!} x^4 + \dots \right] + c_1 \left[x + \frac{1}{3!} x^3 + \frac{1}{5!} x^5 + \dots \right] \\
&= c_0 \sum_{n=0}^{\infty} \frac{1}{(2n)!} x^{2n} + c_1 \sum_{n=0}^{\infty} \frac{1}{(2n+1)!} x^{2n+1} \\
&= c_0 \cosh x + c_1 \sinh x \\
&= c_0 \frac{e^x + e^{-x}}{2} + c_1 \frac{e^x - e^{-x}}{2} \\
&= \left(\frac{c_0 + c_1}{2} \right) e^x + \left(\frac{c_0 - c_1}{2} \right) e^{-x} \\
&= C_0 e^x + C_1 e^{-x}.
\end{aligned}$$

18. Let $y = \sum_{n=0}^{\infty} c_n x^n$. Then

$$y' = \sum_{n=1}^{\infty} n c_n x^{n-1}, \qquad y'' = \sum_{n=2}^{\infty} n(n-1) c_n x^{n-2},$$

and

$$\begin{aligned}
(x+2)y'' + xy' - y &= (x+2) \sum_{n=2}^{\infty} n(n-1) c_n x^{n-2} + x \sum_{n=1}^{\infty} n c_n x^{n-1} - \sum_{n=0}^{\infty} c_n x^n \\
&= \underbrace{\sum_{n=2}^{\infty} n(n-1) c_n x^{n-1}}_{k=n-1} + \underbrace{\sum_{n=2}^{\infty} 2n(n-1) c_n x^{n-2}}_{k=n-2} + \underbrace{\sum_{n=1}^{\infty} n c_n x^n}_{k=n} - \underbrace{\sum_{n=0}^{\infty} c_n x^n}_{k=n} \\
&= \sum_{k=1}^{\infty} (k+1) k c_{k+1} x^k + \sum_{k=0}^{\infty} 2(k+2)(k+1) c_{k+2} x^k + \sum_{k=1}^{\infty} k c_k x^k - \sum_{k=0}^{\infty} c_k x^k \\
&= 2 \cdot 2 \cdot 1 \cdot c_2 x^0 - c_0 x^0 + \sum_{k=1}^{\infty} [(k+1) k c_{k+1} \\
&\qquad + 2(k+2)(k+1) c_{k+2} + (k-1) c_k] x^k \\
&= 4c_2 - c_0 + \sum_{k=1}^{\infty} [(k+1) k c_{k+1} + 2(k+2)(k+1) c_{k+2} + (k-1) c_k] x^k \\
&= 0.
\end{aligned}$$

Thus

$$4c_2 - c_0 = 0$$

and

$$(k+1)kc_{k+1} + 2(k+2)(k+1)c_{k+2} + (k-1)c_k = 0, \qquad k = 1, 2, 3, \ldots$$

or

$$c_2 = \frac{1}{4}c_0$$

and the recurrence relation is

$$c_{k+2} = -\frac{(k+1)kc_{k+1} + (k-1)c_k}{2(k+2)(k+1)}, \qquad k = 1, 2, 3, \ldots .$$

Choose $c_0 \neq 0$ and $c_1 = 0$. Then

$$c_1 = 0$$

$$c_2 = \frac{1}{4}c_0$$

$$c_3 = -\frac{2c_2}{2 \cdot 3 \cdot 2} = -\frac{1}{24}c_0$$

$$c_4 = -\frac{6c_3 + c_2}{24} = -\frac{-c_0/4 + c_0/4}{24} = 0$$

$$c_5 = -\frac{12c_4 + 2c_3}{40} = \frac{1}{480}c_0$$

and so on. Thus, one solution is

$$y_1(x) = c_0\left[1 + \frac{1}{4}x^2 - \frac{1}{24}x^3 + \frac{1}{480}x^5 + \cdots\right].$$

Now choose $c_0 = 0$ and $c_1 \neq 0$. Then

$$c_0 = 0$$

$$c_2 = 0$$

$$c_3 = -\frac{2c_2}{12} = 0$$

$$c_4 = 0$$

and so on. Thus, a second linearly independent solution is

$$y_2(x) = c_1x.$$

24. Let $y = \sum_{n=0}^{\infty} c_n x^n$. Then

$$\begin{aligned}
y'' - xy' - (x+2)y &= \sum_{n=2}^{\infty} n(n-1)c_n x^{n-2} - x\sum_{n=1}^{\infty} nc_n x^{n-1} - (x+2)\sum_{n=0}^{\infty} c_n x^n \\
&= \underbrace{\sum_{n=2}^{\infty} n(n-1)c_n x^{n-2}}_{k=n-2} - \underbrace{\sum_{n=1}^{\infty} nc_n x^n}_{k=n} - \underbrace{\sum_{n=0}^{\infty} c_n x^{n+1}}_{k=n+1} - \underbrace{\sum_{n=0}^{\infty} 2c_n x^n}_{k=n} \\
&= \sum_{k=0}^{\infty}(k+2)(k+1)c_{k+2}x^k - \sum_{k=1}^{\infty} kc_k x^k - \sum_{k=1}^{\infty} c_{k-1}x^k - \sum_{k=0}^{\infty} 2c_k x^k \\
&= 2c_2 - 2c_0 + \sum_{k=1}^{\infty}[(k+2)(k+1)c_{k+2} - (k+2)c_k - c_{k-1}]x^k \\
&= 0.
\end{aligned}$$

Thus

$$2c_2 - 2c_0 = 0$$

$$(k+2)(k+1)c_{k+2} - (k+2)c_k - c_{k-1} = 0, \qquad k = 1, 2, 3, \ldots;$$

or

$$c_2 = c_0$$

$$c_{k+2} = \frac{c_k}{k+1} + \frac{c_{k-1}}{(k+2)(k+1)}, \qquad k = 1, 2, 3, \ldots.$$

Choose $c_0 \neq 0$ and $c_1 = 0$. Then

$$\begin{aligned}
c_1 &= 0 \\
c_2 &= c_0 \\
c_3 &= \frac{c_1}{2} + \frac{c_0}{3\cdot 2} = \frac{1}{3\cdot 2}c_0 \\
c_4 &= \frac{c_2}{3} + \frac{c_1}{4\cdot 3} = \frac{1}{3}c_0 \\
c_5 &= \frac{c_3}{4} + \frac{c_2}{5\cdot 4} = \frac{1}{4\cdot 3\cdot 2}c_0 + \frac{1}{5\cdot 4}c_0 = \frac{5}{5!}c_0 + \frac{3\cdot 2}{5!}c_0 = \frac{11}{5!}c_0,
\end{aligned}$$

and so on. Thus, one solution is

$$y_1(x) = c_0\left[1 + x^2 + \frac{1}{6}x^3 + \frac{1}{3}x^4 + \frac{11}{5!}x^5 + \cdots\right].$$

Now choose $c_0 = 0$ and $c_1 \neq 0$. Then

$$\begin{aligned} c_2 &= c_0 = 0 \\ c_3 &= \frac{c_1}{2} + \frac{c_0}{3 \cdot 2} = \frac{1}{2}c_1 \\ c_4 &= \frac{c_2}{3} + \frac{c_1}{4 \cdot 3} = \frac{1}{4 \cdot 3}c_1 \\ c_5 &= \frac{c_3}{4} + \frac{c_2}{5 \cdot 4} = \frac{1}{4}c_3 = \frac{1}{8}c_1, \end{aligned}$$

and so on. Hence, another solution is

$$y_2(x) = c_1\left[x + \frac{1}{2}x^3 + \frac{1}{12}x^4 + \frac{1}{8}x^5 + \cdots\right].$$

28. Letting $y = \sum_{n=0}^{\infty} c_n x^n$ we have

$$\begin{aligned} (x^2+1)y'' + 2xy' &= (x^2+1)\sum_{n=2}^{\infty} n(n-1)c_n x^{n-2} + 2x\sum_{n=1}^{\infty} nc_n x^{n-1} \\ &= \underbrace{\sum_{n=2}^{\infty} n(n-1)c_n x^n}_{k=n} + \underbrace{\sum_{n=2}^{\infty} n(n-1)c_n x^{n-2}}_{k=n-2} + \underbrace{\sum_{n=1}^{\infty} 2nc_n x^n}_{k=n} \\ &= \sum_{k=2}^{\infty} k(k-1)c_k x^k + \sum_{k=0}^{\infty}(k+2)(k+1)c_{k+2}x^k + \sum_{k=1}^{\infty} 2kc_k x^k \\ &= 2 \cdot 1c_2x^0 + 3 \cdot 2c_3x + 2 \cdot 1c_1x + \sum_{k=2}^{\infty}[k(k-1)c_k + (k+2)(k+1)c_{k+2} + 2kc_k]x^k \\ &= 2c_2 + (6c_3 + 2c_1)x + \sum_{k-2}^{\infty}[(k^2-k)c_k + (k+2)(k+1)c_{k+2} + 2kc_k]x^k \\ &= 2c_2 + (6c_3 + 2c_1)x + \sum_{k=2}^{\infty}[(k^2+k)c_k + (k+2)(k+1)c_{k+2}]x^k \\ &= 2c_2 + (6c_3 + 2c_1)x + \sum_{k=2}^{\infty}[k(k+1)c_k + (k+2)(k+1)c_{k+2}]x^k \\ &= 0. \end{aligned}$$

Thus

$$\begin{aligned} 2c_2 &= 0, \\ 6c_3 + 2c_1 &= 0, \\ k(k+1)c_k + (k+2)(k+1)c_{k+2} &= 0, \qquad k = 2, 3, 4, \ldots, \end{aligned}$$

or

$$c_2 = 0$$

$$c_3 = -\frac{1}{3}c_1$$

and the recurrence relation is

$$c_{k+2} = -\frac{k}{k+2}c_k, \qquad k = 2,\ 3,\ 4,\ \ldots.$$

From $c_2 = 0$ and the recurrence relation we see that $c_{2k} = 0$ for $k = 1,\ 2,\ 3,\ \ldots$. Now

$$c_3 = -\frac{1}{3}c_1$$

$$c_5 = -\frac{3}{5}c_3 = \frac{1}{5}c_1$$

$$c_7 = -\frac{5}{7}c_5 = -\frac{1}{7}c_1$$

$$c_9 = -\frac{7}{9}c_7 = \frac{1}{9}c_9,$$

and so on. The general solution is

$$\begin{aligned} y(x) &= c_0 + c_1 x - \frac{1}{3}c_1 x^3 + \frac{1}{5}c_1 x^5 - \frac{1}{7}c_1 x^7 + \cdots \\ &= c_0 + c_1\left(x - \frac{1}{3}x^3 + \frac{1}{5}x^5 - \frac{1}{7}x^7 + \cdots\right) \\ &= c_0 + c_1 \sum_{n=0}^{\infty} \frac{(-1)^n}{2n+1} x^{2n+1}. \end{aligned}$$

Using $y(0) = 0$ we have

$$c_0 + c_1 \sum_{n=0}^{\infty} \frac{(-1)^n}{2n+1} 0^{2n+1} = c_0 = 0,$$

so

$$y(x) = c_1 \sum_{n=0}^{\infty} \frac{(-1)^n}{2n+1} x^{2n+1}$$

and

$$y'(x) = c_1 \sum_{n=0}^{\infty} (-1)^n x^{2n}.$$

Using $y'(0) = 1$ we have

$$c_1 \sum_{n=0}^{\infty} (-1)^n x^{2n}\bigg|_{x=0} = c_1(1 - x^2 + \ldots)\bigg|_{x=0} = c_1 = 1.$$

Therefore, the solution of the initial value problem is

$$y(x) = \sum_{n=0}^{\infty} \frac{(-1)^n}{2n+1} x^{2n+1}$$
$$= x - \frac{1}{3}x^3 + \frac{1}{5}x^5 - \frac{1}{7}x^7 + \cdots.$$

32. Using $e^x = \sum_{n=0}^{\infty} \frac{1}{n!} x^n = 1 + x + \frac{1}{2}x^2 + \frac{1}{3!}x^3 + \cdots$ we have

$$\begin{aligned}
y'' + e^x y' - y &= \sum_{n=2}^{\infty} n(n-1)c_n x^{n-2} + \left(\sum_{n=0}^{\infty} \frac{1}{n!} x^n\right)\left(\sum_{n=1}^{\infty} n c_n x^{n-1}\right) - \sum_{n=0}^{\infty} c_n x^n \\
&= (2c_2 + 6c_3 x + 12c_4 x^2 + 20c_5 x^3 + \cdots) \\
&\quad + \left(1 + x + \frac{1}{2}x^2 + \frac{1}{6}x^3 + \cdots\right)(c_1 + 2c_2 x + 3c_3 x^2 + \cdots) \\
&\quad - (c_0 + c_1 x + c_2 x^2 + c_3 x^3 + \cdots) \\
&= (2c_2 + 6c_3 x + 12c_4 x^2 + 20c_5 x^3 + \cdots) \\
&\quad + \left[c_1 + (c_1 + 2c_2)x + \left(\frac{1}{2}c_1 + 2c_2 + 3c_3\right)x^2 + \cdots\right] \\
&\quad - (c_0 + c_1 x + c_2 x^2 + \cdots) \\
&= (2c_2 + c_1 - c_0) + (6c_3 + c_1 + 2c_2 - c_1)x \\
&\quad + \left(12c_4 + \frac{1}{2}c_1 + 2c_2 + 3c_3 - c_2\right)x^2 + \cdots \\
&= (2c_2 + c_1 - c_0) + (6c_3 + 2c_2)x + \left(12c_4 + 3c_3 + c_2 + \frac{1}{2}c_1\right)x^2 + \cdots \\
&= 0.
\end{aligned}$$

Thus

$$\begin{aligned}
2c_2 + c_1 - c_0 &= 0 \\
6c_3 + 2c_2 &= 0 \\
12c_4 + 3c_3 + c_2 + \frac{1}{2}c_1 &= 0
\end{aligned}$$

or

$$c_2 = \frac{c_0 - c_1}{2}$$

$$c_3 = -\frac{c_2}{3} = -\frac{c_0 - c_1}{6}$$

$$\begin{aligned} c_4 &= -\frac{3c_3 + c_2 + c_1/2}{12} \\ &= -\frac{6c_3 + 2c_2 + c_1}{24} \\ &= -\frac{-c_0 + c_1 + c_0 - c_1 + c_1}{24} \\ &= -\frac{c_1}{24}, \end{aligned}$$

and so on. The general solution is

$$\begin{aligned} y(x) &= c_0 + c_1 x + \frac{c_0 - c_1}{2}x^2 - \frac{c_0 - c_1}{6}x^3 - \frac{c_1}{24}x^4 + \cdots \\ &= c_0\left[1 + \frac{1}{2}x^2 - \frac{1}{6}x^3 + \cdots\right] + c_1\left[x - \frac{1}{2}x^2 + \frac{1}{6}x^3 - \frac{1}{24}x^4 + \cdots\right], \end{aligned}$$

and two linearly independent solutions are

$$y_1(x) = 1 + \frac{1}{2}x^2 - \frac{1}{6}x^3 + \cdots$$

and

$$y_2(x) = x - \frac{1}{2}x^2 + \frac{1}{6}x^3 - \frac{1}{24}x^4 + \cdots.$$

34. Let $y = \sum_{n=0}^{\infty} c_n x^n$ and recall that $e^x = \sum_{n=0}^{\infty} \frac{1}{n!}x^n$. Then

$$\begin{aligned} y'' - 4xy' - 4y &= \sum_{n=2}^{\infty} n(n-1)c_n x^{n-2} - 4x\sum_{n=1}^{\infty} nc_n x^{n-1} - 4\sum_{n=0}^{\infty} c_n x^n \\ &= \underbrace{\sum_{n=2}^{\infty} n(n-1)c_n x^{n-2}}_{k=n-2} - \underbrace{\sum_{n=1}^{\infty} 4nc_n x^n}_{k=n} - \underbrace{\sum_{n=0}^{\infty} 4c_n x^n}_{k=n} \\ &= \sum_{k=0}^{\infty}(k+2)(k+1)c_{k+2}x^k - \sum_{k=1}^{\infty} 4kc_k x^k - \sum_{k=0}^{\infty} 4c_k x^k \\ &= 2c_2 - 4c_0 + \sum_{k=1}^{\infty}[(k+2)(k+1)c_{k+2} - 4(k+1)c_k]x^k \\ &= e^x \\ &= 1 + \sum_{k=1}^{\infty}\frac{1}{k!}x^k. \end{aligned}$$

Thus

$$2c_2 - 4c_0 = 1,$$

and

$$(k+2)(k+1)c_{k+2} - 4(k+1)c_k = \frac{1}{k!}, \qquad k = 1, 2, 3, \ldots;$$

or

$$c_2 = \frac{1}{2} + 2c_0$$

and the recurrence relation is

$$c_{k+2} = \frac{1}{(k+2)(k+1)k!} + \frac{4c_k}{k+2} = \frac{1}{(k+2)!} + \frac{4}{k+2}c_k, \qquad k = 1, 2, 3, \ldots.$$

Let c_0 and c_1 be arbitrary and iterate to find

$$c_2 = \frac{1}{2} + 2c_0$$

$$c_3 = \frac{1}{3!} + \frac{4}{3}c_1 = \frac{1}{3!} + \frac{4}{3}c_1$$

$$c_4 = \frac{1}{4!} + \frac{4}{4}c_2 = \frac{1}{4!} + \frac{1}{2} + 2c_0 = \frac{13}{4!} + 2c_0$$

$$c_5 = \frac{1}{5!} + \frac{4}{5}c_3 = \frac{1}{5!} + \frac{4}{5\cdot 3!} + \frac{16}{15}c_1 = \frac{17}{5!} + \frac{16}{15}c_1$$

$$c_6 = \frac{1}{6!} + \frac{4}{6}c_4 = \frac{1}{6!} + \frac{4\cdot 13}{6\cdot 4!} + \frac{8}{6}c_0 = \frac{261}{6!} + \frac{4}{3}c_0$$

$$c_7 = \frac{1}{7!} + \frac{4}{7}c_5 = \frac{1}{7!} + \frac{4\cdot 17}{7\cdot 5!} + \frac{64}{105}c_1 = \frac{409}{7!} + \frac{64}{105}c_1.$$

The solution is

$$\begin{aligned} y(x) &= c_0 + c_1x + \left(\frac{1}{2} + 2c_0\right)x^2 + \left(\frac{1}{3!} + \frac{4}{3}c_1\right)x^3 - \left(\frac{13}{4!} + 2c_0\right)x^4 + \left(\frac{17}{5!} + \frac{16}{15}c_1\right)x^5 \\ &\quad + \left(\frac{261}{6!} + \frac{4}{3}c_0\right)x^6 + \left(\frac{409}{7!} + \frac{64}{105}c_1\right)x^7 + \cdots \\ &= c_0\left[1 + 2x^2 + 2x^4 + \frac{4}{3}x^6 + \cdots\right] + c_1\left[x + \frac{4}{3}x^3 + \frac{16}{15}x^5 + \frac{64}{105}x^7 + \cdots\right] \\ &\quad + \frac{1}{2}x^2 + \frac{1}{3!}x^3 + \frac{13}{4!}x^4 + \frac{17}{5!}x^5 + \frac{261}{6!}x^6 + \frac{409}{7!}x^7 + \cdots. \end{aligned}$$

Exercises 6.3

2. Write the differential equation in the form

$$y'' + 0y' - \frac{1}{x(x+3)^2}\,y = 0$$

and identify

$$P(x) = 0 \qquad \text{and} \qquad Q(x) = -\frac{1}{x(x+3)^2}.$$

Then $x = 0$ and $x = -3$ are regular singular points.

8. Write the differential equation in the form

$$y'' + 0y' + \frac{1}{x(x^2+1)^2}\,y = 0$$

and identify

$$P(x) = 0 \qquad \text{and} \qquad Q(x) = \frac{1}{x(x^2+1)^2}.$$

Then $x = 0$, $x = i$, and $x = -i$ are regular singular points.

Throughout the remainder of this section we use the fact that when $y = \sum_{n=0}^{\infty} c_n x^{n+r}$, then

$$y' = \sum_{n=0}^{\infty}(n+r)c_n x^{n+r-1} \qquad \text{and} \qquad y'' = \sum_{n=0}^{\infty}(n+r)(n+r-1)c_n x^{n+r-2}.$$

16. Let $y = \sum_{n=0}^{\infty} c_n x^{n+r}$. Then

$$\begin{aligned}
x^2y'' - \left(x - \frac{2}{9}\right)y &= x^2\sum_{n=0}^{\infty}(n+r)(n+r-1)c_n x^{n+r-2} - \left(x - \frac{2}{9}\right)\sum_{n=0}^{\infty} c_n x^{n+r} \\
&= \sum_{n=0}^{\infty}(n+r)(n+r-1)c_n x^{n+r} - \sum_{n=0}^{\infty} c_n x^{n+r+1} + \sum_{n=0}^{\infty}\frac{2}{9}c_n x^{n+r} \\
&= x^r\Bigg[\underbrace{\sum_{n=0}^{\infty}(n+r)(n+r-1)c_n x^n}_{k=n} - \underbrace{\sum_{n=0}^{\infty} c_n x^{n+1}}_{k=n+1} + \underbrace{\sum_{n=0}^{\infty}\frac{2}{9}c_n x^n}_{k=n}\Bigg] \\
&= x^r\left[\sum_{k=0}^{\infty}(k+r)(k+r-1)c_k x^k - \sum_{k=1}^{\infty} c_{k-1}x^k + \sum_{k=0}^{\infty}\frac{2}{9}c_k x^k\right] \\
&= x^r\left[r(r-1)c_0 + \frac{2}{9}c_0 + \sum_{k=1}^{\infty}\left((k+r)(k+r-1)c_k - c_{k-1} + \frac{2}{9}c_k\right)x^k\right] \\
&= x^r\left[\left(r^2 - r + \frac{2}{9}\right)c_0 + \sum_{k=1}^{\infty}\left(\left[(k+r)(k+r-1) + \frac{2}{9}\right]c_k - c_{k-1}\right)x^k\right] \\
&= 0.
\end{aligned}$$

The indicial equation is

$$r^2 - r + \frac{2}{9} = \left(r - \frac{2}{3}\right)\left(r - \frac{1}{3}\right) = 0.$$

Thus, $r_1 = 2/3$ and $r_2 = 1/3$. Note that the difference of the roots $r_1 - r_2 = 1/3$ is not an integer. For $r_1 = 2/3$ we obtain the recurrence relation

$$\left[\left(k + \frac{2}{3}\right)\left(k + \frac{2}{3} - 1\right) + \frac{2}{9}\right]c_k = c_{k-1}$$

or

$$c_k = \frac{3c_{k-1}}{3k^2 + k}, \qquad k = 1, 2, 3, \ldots.$$

If we take c_0 to be arbitrary and iterate, then

$$c_1 = \frac{3}{4}c_0$$
$$c_2 = \frac{3}{14}c_1 = \frac{9}{56}c_0$$
$$c_3 = \frac{1}{10}c_2 = \frac{9}{560}c_0$$
$$c_4 = \frac{3}{52}c_3 = \frac{27}{29120}c_0,$$

and so on. Thus, one solution is

$$\begin{aligned} y_1(x) &= x^{2/3}[c_0 + c_1 x + c_2 x^2 + c_3 x^3 + c_4 x^4 + \cdots] \\ &= c_0 x^{2/3}\left[1 + \frac{3}{4}x + \frac{9}{56}x^2 + \frac{9}{560}x^3 + \frac{27}{29120}x^4 + \cdots\right]. \end{aligned}$$

For $r_2 = 1/3$ we obtain the recurrence relation

$$\left[\left(k + \frac{1}{3}\right)\left(k + \frac{1}{3} - 1\right) + \frac{2}{9}\right]c_k = c_{k-1}$$

or

$$c_k = \frac{3c_{k-1}}{3k^2 - k}, \qquad k = 1, 2, 3, \ldots.$$

If we take c_0 to be arbitrary and iterate, then

$$c_1 = \frac{3}{2}c_0$$
$$c_2 = \frac{3}{10}c_1 = \frac{9}{20}c_0$$
$$c_3 = \frac{1}{8}c_2 = \frac{9}{160}c_0$$
$$c_4 = \frac{3}{44}c_3 = \frac{27}{7040}c_0,$$

and so on. Thus, a second solution is

$$y_2(x) = c_0 x^{1/3}\left[1 + \frac{3}{2}x + \frac{9}{20}x^2 + \frac{9}{160}x^3 + \frac{27}{7040}x^4 + \cdots\right].$$

The general solution is

$$y(x) = C_1 x^{2/3}\left[1 + \frac{3}{4}x + \frac{9}{56}x^2 + \frac{9}{560}x^3 + \frac{27}{29120}x^4 + \cdots\right]$$
$$+ C_2 x^{1/3}\left[1 + \frac{3}{2}x + \frac{9}{20}x^2 + \frac{9}{160}x^3 + \frac{27}{7040}x^4 + \cdots\right].$$

22. Let $y = \sum_{n=0}^{\infty} c_n x^{n+r}$. Then

$$x(x-2)y'' + y' - 2y$$
$$= (x^2 - 2x)\sum_{n=0}^{\infty}(n+r)(n+r-1)c_n x^{n+r-2} + \sum_{n=0}^{\infty}(n+r)c_n x^{n+r-1} - \sum_{n=0}^{\infty} 2c_n x^{n+r}$$
$$= \sum_{n=0}^{\infty}(n+r)(n+r-1)c_n x^{n+r} - \sum_{n=0}^{\infty} 2(n+r)(n+r-1)c_n x^{n+r-1}$$
$$+ \sum_{n=0}^{\infty}(n+r)c_n x^{n+r+-1} - \sum_{n=0}^{\infty} 2c_n x^{n+r}$$
$$= x^r\left[\sum_{n=0}^{\infty}\Big[(n+r)(n+r-1) - 2\Big]c_n x^n - \sum_{n=0}^{\infty}\Big[2(n+r)(n+r-1) - (n+r)\Big]c_n x^{n-1}\right]$$
$$= x^r\Bigg[\underbrace{\sum_{n=0}^{\infty}\Big[(n+r)(n+r-1) - 2\Big]c_n x^n}_{k=n} - \underbrace{\sum_{n=0}^{\infty}(n+r)(2n+2r-3)c_n x^{n-1}}_{k=n-1}\Bigg]$$
$$= x^r\left[\sum_{k=0}^{\infty}\Big[(k+r)(k+r-1) - 2\Big]c_k x^k - \sum_{k=-1}^{\infty}(k+r+1)(2k+2r-1)c_{k+1}x^k\right]$$
$$= x^r\Bigg[-r(2r-3)c_0 x^{-1} + \sum_{k=0}^{\infty}\bigg(\Big[(k+r)(k+r-1) - 2\Big]c_k$$
$$- (k+r+1)(2k+2r-1)c_{k+1}\bigg)x^k\Bigg]$$
$$= 0.$$

The indicial equation is $r(2r-3) = 0$ and the roots are $r_1 = 3/2$ and $r_2 = 0$. The recurrence relation is

$$c_{k+1} = \frac{(k+r)(k+r-1) - 2}{(k+r+1)(2k+2r-1)}c_k, \qquad k = 0, 1, 2, \ldots.$$

For $r_1 = 3/2$ the recurrence relation is

$$c_{k+1} = \frac{\left(k+\frac{3}{2}\right)\left(k+\frac{1}{2}\right)-2}{\left(k+\frac{5}{2}\right)(2k+2)}c_k$$

$$= \frac{(2k+3)(2k+1)-8}{(2k+5)(4k+4)}c_k$$

$$= \frac{4k^2+8k-5}{(2k+5)(4k+4)}c_k$$

$$= \frac{(2k+5)(2k-1)}{(2k+5)(4k+4)}c_k$$

$$= \frac{2k-1}{4(k+1)}c_k, \qquad k = 0,\, 1,\, 2,\, \ldots .$$

Then

$$c_1 = -\frac{1}{4}c_0$$

$$c_2 = \frac{1}{8}c_1 = -\frac{1}{32}c_0$$

$$c_3 = \frac{3}{12}c_2 = -\frac{1}{128}c_0$$

$$c_4 = \frac{5}{16}c_3 = -\frac{5}{2048}c_0,$$

and so on. Thus, one solution is

$$y_1(x) = c_0 x^{3/2}\left[1 - \frac{1}{4}x - \frac{1}{32}x^2 - \frac{1}{128}x^3 - \frac{5}{2048}x^4 - \cdots\right].$$

To find a second linearly independent solution we use $r_2 = 0$. The recurrence relation is

$$c_{k+1} = \frac{k(k-1)-2}{(k+1)(2k-1)}c_k$$

$$= \frac{k^2-k-2}{(k+1)(2k-1)}c_k$$

$$= \frac{(k+1)(k-2)}{(k+1)(2k-1)}c_k$$

$$= \frac{k-2}{2k-1}c_k, \qquad k = 0,\, 1,\, 2,\, \ldots .$$

Then

$$\begin{aligned} c_1 &= 2c_0 \\ c_2 &= \frac{-1}{1}c_1 = -2c_0 \\ c_3 &= 0 \\ c_4 &- 0, \end{aligned}$$

and so on. A second linearly independent solution is

$$y_2(x) = c_0[1 + 2x - 2x^2].$$

The general solution is

$$y(x) = C_1 x^{3/2}\left[1 - \frac{1}{4}x - \frac{1}{32}x^2 - \frac{1}{128}x^3 - \cdots\right] + C_2[1 + 2x - 2x^2].$$

28. Let $y = \sum_{n=0}^{\infty} c_n x^{n+r}$. Then

$$\begin{aligned} xy'' + y &= x\sum_{n=0}^{\infty}(n+r)(n+r-1)c_n x^{n+r-2} + \sum_{n=0}^{\infty} c_n x^{n+r} \\ &= \sum_{n=0}^{\infty}(n+r)(n+r-1)c_n x^{n+r-1} + \sum_{n=0}^{\infty} c_n x^{n+r} \\ &= x^r\Bigg[\underbrace{\sum_{n=0}^{\infty}(n+r)(n+r-1)c_n x^{n-1}}_{k=n-1} + \underbrace{\sum_{n=0}^{\infty} c_n x^n}_{k=n}\Bigg] \\ &= x^r\left[\sum_{k=-1}^{\infty}(k+r+1)(k+r)c_{k+1}x^k + \sum_{k=0}^{\infty} c_k x^k\right] \\ &= x^r\left[r(r-1)c_0 x^{-1} + \sum_{k=0}^{\infty}[(k+r+1)(k+r)c_{k+1} + c_k]x^k\right] \\ &= 0. \end{aligned}$$

The indicial equation is $r(r-1) = 0$, so $r_1 = 1$ and $r_2 = 0$. The difference of the roots $r_1 - r_2 = 1 - 0 = 1$ is an integer. For $r_1 = 1$ we obtain the recurrence relation

$$c_{k+1} = \frac{-c_k}{(k+2)(k+1)}, \qquad k = 0, 1, 2, \ldots .$$

Then

$$c_1 = -\frac{1}{2}c_0$$
$$c_2 = -\frac{1}{3 \cdot 2}c_1 = \frac{1}{3! \cdot 2}c_0$$
$$c_3 = \frac{1}{4 \cdot 3}c_2 = -\frac{1}{4!3!}c_0$$
$$c_4 = -\frac{1}{5 \cdot 4}c_3 = \frac{1}{5!4!}c_0,$$

and so on. Thus, one solution is

$$y_1(x) = x\left[c_0 - \frac{1}{2}c_0x + \frac{1}{3!2}c_0x^2 - \frac{1}{4!3!}c_0x^3 + \frac{1}{5!4!}c_0x^4 - \dots\right]$$
$$= c_0\left[x - \frac{1}{2}x^2 + \frac{1}{3!2}x^3 - \frac{1}{4!3!}x^4 + \frac{1}{5!4!}x^5 - \dots\right]$$
$$= c_0 \sum_{n=1}^{\infty} \frac{(-1)^{n+1}}{n!(n-1)!}x^n.$$

To find a second linearly independent solution we use formula (35):

$$y_2 = y_1 \int \frac{e^{\int -0\,dx}}{y_1^2}\,dx = y_1 \int \frac{1}{y_1^2}\,dx$$
$$= y_1 \int \frac{dx}{\left(x - \frac{1}{2}x^2 + \frac{1}{3!2}x^3 - \frac{1}{4!3!}x^4 + \dots\right)}$$
$$= y_1 \int \frac{dx}{x^2 - x^3 + \frac{5}{3!2}x^4 - \frac{14}{4!3!}x^5 + \dots}$$
$$= y_1 \int \frac{dx}{x^2\left(1 - x + \frac{5}{3!2}x^2 - \frac{14}{4!3!}x^3 + \dots\right)}$$
$$= y_1 \int \frac{1}{x^2}\left(1 + x + \frac{7}{3!2}x^2 + \frac{38}{4!3!}x^3 + \dots\right)dx$$
$$= y_1 \int \left(\frac{1}{x^2} + \frac{1}{x} + \frac{7}{3!2} + \frac{38}{4!3!}x + \dots\right)dx$$
$$= y_1\left(-\frac{1}{x} + \ln x + \frac{7}{3!2}x + \frac{19}{4!3!}x^2 + \dots\right).$$

The general solution is

$$y_1(x) = C_1y_1 + C_2y_1\left(-\frac{1}{x} + \ln x + \frac{7}{3!2}x + \frac{19}{4!3!}x^2 + \dots\right).$$

36. Let $y = \sum_{n=0}^{\infty} c_n x^{n+r}$. Then

$$\begin{aligned}
x^2y'' - y' + y &= x^2 \sum_{n=0}^{\infty}(n+r)(n+r-1)c_n x^{n+r-2} - \sum_{n=0}^{\infty}(n+r)c_n x^{n+r-1} + \sum_{n=0}^{\infty} c_n x^{n+r} \\
&= \underbrace{\sum_{n=0}^{\infty}(n+r)(n+r-1)c_n x^{n+r}}_{k=n} - \underbrace{\sum_{n=0}^{\infty}(n+r)c_n x^{n+r-1}}_{k=n-1} + \underbrace{\sum_{n=0}^{\infty} c_n x^{n+r}}_{k=n} \\
&= x^r\left[\sum_{k=0}^{\infty}(k+r)(k+r-1)c_k x^k - \sum_{k=-1}^{\infty}(k+r+1)c_{k+1}x^k + \sum_{k=0}^{\infty} c_k x^k\right] \\
&= x^r\left[rc_0x^{-1} + \sum_{k=0}^{\infty}\Big(\left[(k+r)(k+r-1)+1\right]c_k - (k+r+1)c_{k+1}\Big)x^k\right] \\
&= 0.
\end{aligned}$$

Thus $r = 0$ and the recurrence relation is

$$c_{k+1} = \frac{k(k-1)+1}{k+1}c_k, \qquad k = 0,\ 1,\ 2,\ \dots .$$

Then

$$\begin{aligned}
c_1 &= 0 \\
c_2 &= \frac{1}{2}c_1 = \frac{1}{2}c_0 \\
c_3 &= c_2 = \frac{1}{2}c_0 \\
c_4 &= \frac{7}{4}c_3 = \frac{7}{8}c_0,
\end{aligned}$$

and so on. Therefore, one solution is

$$\begin{aligned}
y(x) &= c_0 + c_0x + \frac{1}{2}c_0x^2 + \frac{1}{2}c_0x^3 + \frac{7}{8}c_0x^4 + \dots \\
&= c_0\left[1 + x + \frac{1}{2}x^2 + \frac{1}{2}x^3 + \frac{7}{8}x^4 + \dots\right].
\end{aligned}$$

Exercises 6.4

4. Write the equation in the form

$$x^2y'' + xy' + \left(x^2 - \frac{1}{16}\right)y = 0.$$

This is Bessel's equation with $\nu = 1/4$. From (9) in the text, the general solution is

$$y(x) = c_1 J_{1/4}(x) + c_2 J_{-1/4}(x).$$

10. From $y = x^n J_n(x)$ we find

$$y' = x^n J_n' + nx^{n-1}J_n \qquad \text{and} \qquad y'' = x^n J_n'' + 2nx^{n-1}J_n' + n(n-1)x^{n-2}J_n.$$

Substituting into the differential equation, we have

$$\begin{aligned}
x^{n+1}J_n'' &+ 2nx^nJ_n' + n(n-1)x^{n-1}J_n + (1-2n)(x^nJ_n' + nx^{n-1}J_n) + x^{n+1}J_n \\
&= x^{n+1}J_n'' + (2n+1-2n)x^nJ_n' + (n^2 - n + n - 2n^2)x^{n-1}J_n + x^{n+1}J_n \\
&= x^{n+1}[x^2J_n'' + xJ_n' - n^2J_n + x^2J_n] \\
&= x^{n+1}[x^2J_n'' + xJ_n' + (x^2 - n^2)J_n] \\
&= x^{n-1}\cdot 0 \qquad \text{(since } J_n \text{ is a solution of Bessel's equation)} \\
&= 0.
\end{aligned}$$

Therefore, x^nJ_n is a solution of the original equation.

18. For the differential equation

$$xy'' - 5xy' + xy = 0$$

identify $n = 3$. Then by Problem 10 a solution is $x^3J_3(x)$.

22. Using

$$J_\nu(x) = \sum_{n=0}^{\infty} \frac{(-1)^n}{n!\Gamma(1+\nu+n)}\left(\frac{x}{2}\right)^{2n+\nu}$$

$$J_\nu'(x) = \sum_{n=0}^{\infty} \frac{(2n+\nu)(-1)^n}{2n!\Gamma(1+\nu+n)}\left(\frac{x}{2}\right)^{2n+\nu-1}$$

$$J_{\nu-1}(x) = \sum_{n=0}^{\infty} \frac{(-1)^n}{n!\Gamma(\nu+n)}\left(\frac{x}{2}\right)^{2n+\nu-1}$$

we obtain

$$\begin{aligned}
\frac{d}{dx}[x^\nu J_\nu(x)] &= x^\nu J_\nu'(x) + \nu x^{\nu-1} J_\nu(x) \\
&= x^\nu \sum_{n=0}^{\infty} \frac{(2n+\nu)(-1)^n}{2n!\Gamma(1+\nu+n)} \left(\frac{x}{2}\right)^{2n+\nu-1} \\
&\qquad + \nu x^{\nu-1} \sum_{n=0}^{\infty} \frac{(-1)^n}{n!\Gamma(1+\nu+n)} \left(\frac{x}{2}\right)^{2n+\nu} \\
&= x^\nu \sum_{n=0}^{\infty} \frac{(2n+\nu)(-1)^n}{2n!(\nu+n)\Gamma(\nu+n)} \left(\frac{x}{2}\right)^{2n+\nu-1} \\
&\qquad + x^\nu \sum_{n=0}^{\infty} \frac{\nu(-1)^n 2^{-1}}{n!(\nu+n)\Gamma(\nu+n)} \left(\frac{x}{2}\right)^{-1} \left(\frac{x}{2}\right)^{2n+\nu} \\
&= x^\nu \left[\sum_{n=0}^{\infty} \frac{(2n+\nu)(-1)^n}{2n!(\nu+n)\Gamma(\nu+n)} \left(\frac{x}{2}\right)^{2n+\nu-1} \right. \\
&\qquad \left. + \sum_{n=0}^{\infty} \frac{\nu(-1)^n}{2n!(\nu+n)\Gamma(\nu+n)} \left(\frac{x}{2}\right)^{2n+\nu-1} \right] \\
&= x^\nu \sum_{n=0}^{\infty} \frac{(2n+2\nu)(-1)^n}{2n!(\nu+n)\Gamma(\nu+n)} \left(\frac{x}{2}\right)^{2n+\nu-1} \\
&= x^\nu \sum_{n=0}^{\infty} \frac{(-1)^n}{n!\Gamma(\nu+n)} \left(\frac{x}{2}\right)^{2n+\nu-1} \\
&= x^\nu J_{\nu-1}(x).
\end{aligned}$$

Alternatively, we can note that the formula in Problem 19 is a linear first-order differential equation in $J_\nu(x)$. An integrating factor for this equation is x^ν, so

$$\frac{d}{dx}[x^\nu J_\nu(x)] = x^\nu J_{\nu-1}(x).$$

26. As noted in Problem 22,

$$\int x^\nu J_{\nu-1}(x)\,dx = x^\nu J_\nu(x).$$

Now, integrating by parts, we have

$$\int x^3 J_0(x)\,dx = \int x^2 (xJ_0(x)\,dx)$$

$u = x^2$	$dv = xJ_0(x)\,dx$
$du = 2x\,dx$	$v = xJ_1(x)$

$$\begin{aligned}
&= x^3 J_1(x) - \int 2x^2 J_1(x)\,dx \\
&= x^3 J_1(x) - 2\int x^2 J_1(x)\,dx \\
&= x^3 J_1(x) - 2x^2 J_2(x) + c.
\end{aligned}$$

30. From Problem 21 we have

$$J_{\nu+1}(x) = \frac{2\nu}{x} J_\nu(x) - J_{\nu-1}(x)$$

and from Problem 28 we have

$$J_{3/2}(x) = \sqrt{\frac{2}{\pi x}}\left(\frac{\sin x}{x} - \cos x\right).$$

Thus, using $\nu = 3/2$ and Problem 27,

$$\begin{aligned}
J_{5/2}(x) &= \frac{3}{x} J_{3/2}(x) - J_{1/2}(x) \\
&= \frac{3}{x}\sqrt{\frac{2}{\pi x}}\left(\frac{1}{x}\sin x - \cos x\right) - \sqrt{\frac{2}{\pi x}}\sin x \\
&= \sqrt{\frac{2}{\pi x}}\left[\left(\frac{3}{x^2} - 1\right)\sin x - \frac{3}{x}\cos x\right].
\end{aligned}$$

40. We use the product rule for differentiation:

$$\begin{aligned}
\frac{d}{dx}\left[(1-x^2)\frac{dy}{dx}\right] + n(n+1)y &= (1-x^2)\frac{d^2y}{dx^2} + (-2x)\frac{dy}{dx} + n(n+1)y \\
&= (1-x^2)y'' - 2xy' + n(n+1)y = 0.
\end{aligned}$$

44. Letting $x = 1$ in $(1 - 2xt + t^2)^{-1/2}$, we have

$$\begin{aligned}
(1 - 2t + t^2)^{-1/2} &= (1-t)^{-1} \\
&= \frac{1}{1-t} \\
&= 1 + t + t^2 + t^3 + \dots \qquad (|t| < 1) \\
&= \sum_{n=0}^{\infty} t^n.
\end{aligned}$$

From Problem 43 we have

$$\sum_{n=0}^{\infty} P_n(1) t^n = (1 - 2t + t^2)^{-1/2} = \sum_{n=0}^{\infty} t^n.$$

Equating the coefficients of corresponding terms in the two series, we see that $P_n(1) = 1$. Similarly, letting $x = -1$ we have

$$\begin{aligned}
(1 + 2t + t^2)^{-1/2} &= (1 + t)^{-1} \\
&= \frac{1}{1+t} \\
&= 1 - t + t^2 - 3t^3 + \dots \qquad (|t| < 1) \\
&= \sum_{n=0}^{\infty} (-1)^n t^n \\
&= \sum_{n=0}^{\infty} P_n(-1) t^n,
\end{aligned}$$

so that $P_n(-1) = (-1)^n$.

7 The Laplace Transform

Exercises 7.1

4. To find the Laplace transform of a piecewise defined function we integrate separately over the two intervals on which the function is defined:

$$\begin{aligned}
\mathscr{L}\{f(t)\} &= \int_0^\infty e^{-st} f(t)\,dt \\
&= \int_0^1 e^{-st} f(t)\,dt + \int_1^\infty e^{-st} f(t)\,dt \\
&= \int_0^1 e^{-st}(2t+1)\,dt + \int_1^\infty e^{-st}\cdot 0\,dt
\end{aligned}$$

$u = 2t+1$	$dv = e^{-st}\,dt$
$du = 2\,dt$	$v = -\frac{1}{s}e^{-st}$

$$\begin{aligned}
&= -\frac{e^{-st}}{s}(2t+1)\Big|_0^1 + \frac{2}{s}\int_0^1 e^{-st}\,dt \\
&= -\frac{3e^{-s}}{s} + \frac{1}{s} - \frac{2e^{-s}}{s^2} + \frac{2}{s^2}.
\end{aligned}$$

14. We use integration by parts twice, together with the transforms of $\cos t$ and $\sin t$:

$$\mathscr{L}\{t\sin t\} = \int_0^\infty e^{-st} t\sin t\,dt$$

$u = t\sin t$	$dv = e^{-st}\,dt$
$du = (t\cos t + \sin t)\,dt$	$v = -\frac{1}{s}e^{-st}$

$$\begin{aligned}
&= -\frac{e^{-st} t\sin t}{s}\Big|_0^\infty + \frac{1}{s}\int_0^\infty e^{-st}(t\cos t + \sin t)\,dt \\
&= \frac{1}{s}\int_0^\infty e^{-st} t\cos t\,dt + \frac{1}{s}\int_0^\infty e^{-st}\sin t\,dt
\end{aligned}$$

$u = t\cos t$	$dv = e^{-st}\,dt$
$du = (-t\sin t + \cos t)\,dt$	$v = -\frac{1}{s}e^{-st}$

$$= \frac{1}{s}\left[-\frac{e^{-st}t\cos t}{s}\Big|_0^\infty + \frac{1}{s}\int_0^\infty e^{-st}(-t\sin t + \cos t)\,dt\right] + \frac{1}{s}\int_0^\infty e^{-st}\sin t\,dt$$

$$= -\frac{1}{s^2}\int_0^\infty e^{-st}t\sin t\,dt + \frac{1}{s^2}\int_0^\infty e^{-st}\cos t\,dt + \frac{1}{s}\int_0^\infty e^{-st}\sin t\,dt$$

$$= -\frac{1}{s^2}\mathscr{L}\{t\sin t\} + \frac{1}{s^2}\mathscr{L}\{\cos t\} + \frac{1}{s}\mathscr{L}\{\sin t\}.$$

Solving for $\mathscr{L}\{t\sin t\}$ gives

$$\mathscr{L}\{t\sin t\} = \frac{s^2}{s^2+1}\left[\frac{1}{s^2}\mathscr{L}\{\cos t\} + \frac{1}{s}\mathscr{L}\{\sin t\}\right]$$

$$= \frac{s^2}{s^2+1}\left[\frac{1}{s^2}\cdot\frac{s}{s^2+1} + \frac{1}{s}\cdot\frac{1}{s^2+1}\right]$$

$$= \frac{2s}{(s^2+1)^2}.$$

22. We expand $(2t-1)^3$ and use Theorem 7.2:

$$\mathscr{L}\left\{(2t-1)^3\right\} = \mathscr{L}\left\{8t^3 - 12t^2 + 6t - 1\right\}$$

$$= 8\,\mathscr{L}\left\{t^3\right\} - 12\,\mathscr{L}\left\{t^2\right\} + 6\,\mathscr{L}\{t\} - \mathscr{L}\{1\}$$

$$= 8\cdot\frac{3!}{s^4} - 12\cdot\frac{2!}{s^3} + 6\cdot\frac{1}{s^2} - \frac{1}{s}$$

$$= \frac{48}{s^4} - \frac{24}{s^3} + \frac{6}{s^2} - \frac{1}{s}.$$

34. Recall from trigonometry that

$$\cos^2 t = \frac{1}{2}(1 + \cos 2t).$$

Hence

$$\mathscr{L}\left\{\cos^2 t\right\} = \frac{1}{2}\mathscr{L}\{1\} + \frac{1}{2}\mathscr{L}\{\cos 2t\}$$

$$= \frac{1}{2}\frac{1}{s} + \frac{1}{2}\frac{s}{s^2+4}$$

$$= \frac{s^2+2}{s(s^2+4)}.$$

36. We use the trigonometric identity

$$\sin A\sin B = \frac{1}{2}\left[\cos(A-B) - \cos(A+B)\right].$$

With $A = t$ and $B = 2t$ we have

$$\sin t \sin 2t = \frac{1}{2}\left[\cos(-t) - \cos 3t\right] = \frac{1}{2}\left[\cos t - \cos 3t\right].$$

Then

$$\mathscr{L}\{\sin t \sin 2t\} = \frac{1}{2}\left[\mathscr{L}\{\cos t\} - \mathscr{L}\{\cos 3t\}\right] = \frac{1}{2}\left[\frac{s}{s^2+1} - \frac{s}{s^2+9}\right].$$

42. Identifying $\alpha = 3/2$ in Problem 39 and the fact that $\Gamma(1/2) = \sqrt{\pi}$ we have

$$\mathscr{L}\left\{t^{3/2}\right\} = \frac{\Gamma(5/2)}{s^{5/2}} = \frac{3}{2}\Gamma(3/2)\frac{1}{s^{5/2}} = \frac{3}{2}\cdot\frac{1}{2}\Gamma(1/2)\frac{1}{s^{5/2}} = \frac{3\sqrt{\pi}}{4s^{5/2}}.$$

Exercises 7.2

4. We expand $(s+2)^2$ and divide:

$$\begin{aligned}\mathscr{L}^{-1}\left\{\frac{(s+2)^2}{s^3}\right\} &= \mathscr{L}^{-1}\left\{\frac{s^2+4s+4}{s^3}\right\}\\ &= \mathscr{L}^{-1}\left\{\frac{1}{s} + \frac{4}{s^2} + \frac{4}{s^3}\right\}\\ &= \mathscr{L}^{-1}\left\{\frac{1}{s}\right\} + 4\mathscr{L}^{-1}\left\{\frac{1}{s^2}\right\} + \frac{4}{2}\mathscr{L}^{-1}\left\{\frac{2!}{s^3}\right\}\\ &= 1 + 4t + 2t^2.\end{aligned}$$

8. We factor 5 out of the denominator:

$$\mathscr{L}^{-1}\left\{\frac{1}{5s-2}\right\} = \frac{1}{5}\mathscr{L}^{-1}\left\{\frac{1}{s-2/5}\right\} = \frac{1}{5}e^{2t/5}.$$

22. Using partial fractions we can write

$$\begin{aligned}\mathscr{L}\left\{\frac{s+1}{(s^2-4s)(s+5)}\right\} &= \mathscr{L}^{-1}\left\{\frac{s+1}{s(s-4)(s+5)}\right\}\\ &= \mathscr{L}^{-1}\left\{-\frac{1/20}{s} + \frac{5/36}{s-4} - \frac{4/45}{s+5}\right\}\\ &= -\frac{1}{20}\mathscr{L}^{-1}\left\{\frac{1}{s}\right\} + \frac{5}{36}\mathscr{L}^{-1}\left\{\frac{1}{s-4}\right\} - \frac{4}{45}\mathscr{L}^{-1}\left\{\frac{1}{s+5}\right\}\\ &= -\frac{1}{20} + \frac{5}{36}e^{4t} - \frac{4}{45}e^{-5t}.\end{aligned}$$

26. Using partial fractions and Theorem 7.3 we have

$$\begin{aligned}\mathscr{L}^{-1}\left\{\frac{1}{s^4-9}\right\} &= \mathscr{L}^{-1}\left\{\frac{1/6}{s^2-3} - \frac{1/6}{s^2+3}\right\}\\ &= \frac{1}{6}\mathscr{L}^{-1}\left\{\frac{1}{s^2-3}\right\} - \frac{1}{6}\mathscr{L}^{-1}\left\{\frac{1}{s^2+3}\right\}\\ &= \frac{1}{6\sqrt{3}}\mathscr{L}^{-1}\left\{\frac{\sqrt{3}}{s^2-3}\right\} - \frac{1}{6\sqrt{3}}\mathscr{L}^{-1}\left\{\frac{\sqrt{3}}{s^2+3}\right\}\\ &= \frac{1}{6\sqrt{3}}\sinh\sqrt{3}\,t - \frac{1}{6\sqrt{3}}\sin\sqrt{3}\,t.\end{aligned}$$

Exercises 7.3

12. Using

$$\cos^2 3t = \frac{1}{2}(1+\cos 6t)$$

we have

$$\begin{aligned}\mathscr{L}\left\{\cos^2 3t\right\} &= \frac{1}{2}\mathscr{L}\{1\} + \frac{1}{2}\mathscr{L}\{\cos 6t\}\\ &= \frac{1/2}{s} + \frac{s/2}{s^2+36}\\ &= \frac{s^2+18}{s(s^2+36)}.\end{aligned}$$

Then, using the first translation theorem,

$$\mathscr{L}\left\{e^t\cos^2 3t\right\} = \frac{(s-1)^2+18}{(s-1)[(s-1)^2+36]}.$$

18. Completing the square in the denominator, we have

$$\begin{aligned}\mathscr{L}^{-1}\left\{\frac{2s+5}{s^2+6s+34}\right\} &= \mathscr{L}^{-1}\left\{\frac{2s+5}{(s+3)^2+25}\right\}\\ &= \mathscr{L}^{-1}\left\{\frac{2s+6-1}{(s+3)^2+25}\right\}\\ &= 2\mathscr{L}^{-1}\left\{\frac{s+3}{(s+3)^2+25}\right\} - \frac{1}{5}\mathscr{L}^{-1}\left\{\frac{5}{(s+3)^2+25}\right\}\\ &= 2e^{-3t}\cos 5t - \frac{1}{5}e^{-3t}\sin 5t.\end{aligned}$$

24. Expressing e^{2-t} in the form $f(t-2)$ we obtain

$$\mathscr{L}\left\{e^{2-t}\mathscr{U}(t-2)\right\} = \mathscr{L}\left\{e^{-(t-2)}\mathscr{U}(t-2)\right\}$$
$$= e^{-2s}\frac{1}{s+1}.$$

36. Use partial fractions to write

$$\frac{1}{s^2(s-1)} = -\frac{1}{s^2} - \frac{1}{s} + \frac{1}{s-1}.$$

Then

$$\mathscr{L}^{-1}\left\{\frac{e^{-2s}}{s^2(s-1)}\right\} = -\mathscr{L}^{-1}\left\{\frac{e^{-2s}}{s^2}\right\} - \mathscr{L}^{-1}\left\{\frac{e^{-2s}}{s}\right\} + \mathscr{L}^{-1}\left\{\frac{e^{-2s}}{s-1}\right\}$$
$$= \left[-(t-2) - 1 + e^{t-2}\right]\mathscr{U}(t-2)$$
$$= \left(e^{t-2} - t + 1\right)\mathscr{U}(t-2).$$

42. We use Theorem 7.7 followed by Theorem 7.5:

$$\mathscr{L}\left\{te^{-3t}\cos 3t\right\} = -\frac{d}{ds}\left(e^{-3t}\cos 3t\right)$$
$$= -\frac{d}{ds}\mathscr{L}\{\cos 3t\}_{s\to s+3}$$
$$= -\frac{d}{ds}\left[\frac{s+3}{(s+3)^2+9}\right]$$
$$= \frac{(s+3)^2-9}{[(s+3)^2+9]^2}.$$

48. We use the trigonometric identity

$$\cos\left(t - \frac{3\pi}{2}\right) = -\sin t.$$

Then

$$f(t) = \sin t\,\mathscr{U}\left(t - \frac{3\pi}{2}\right)$$
$$= -\cos\left(t - \frac{3\pi}{2}\right)\mathscr{U}\left(t - \frac{3\pi}{2}\right)$$

so

$$\mathscr{L}\{f\} = -\mathscr{L}\left\{\cos\left(t - \frac{3\pi}{2}\right)\mathscr{U}\left(t - \frac{3\pi}{2}\right)\right\}$$
$$= -e^{-3\pi s/2}\frac{s}{s^2+1}.$$

52. From the graph we identify

$$f(t) = \mathscr{U}(t-1) + \mathscr{U}(t-2) + \mathscr{U}(t-3) + \dots.$$

Then

$$\mathscr{L}\{f\} = \frac{1}{s}\left[e^{-s} + e^{-2s} + e^{-3s} + \dots\right].$$

Since the sum is an infinite geometric progression with common ratio e^{-s}, we can write

$$\mathscr{L}\{f\} = \frac{1}{s}\frac{e^{-s}}{1-e^{-s}}.$$

56. Using the alternate form of Theorem 7.7 for $n = 1$ we have

$$\begin{aligned}
\mathscr{L}^{-1}\left\{\ln\frac{s^2+1}{s^2+4}\right\} &= \mathscr{L}^{-1}\{F(s)\} \\
&= -\frac{1}{t}\mathscr{L}^{-1}\left\{\frac{d}{ds}F(s)\right\} \\
&= -\frac{1}{t}\mathscr{L}^{-1}\left\{\frac{d}{ds}\ln\frac{s^2+1}{s^2+4}\right\} \\
&= -\frac{1}{t}\mathscr{L}^{-1}\left\{\frac{s^2+4}{s^2+1}\cdot\frac{(s^2+4)2s-(s^2+1)2s}{(s^2+4)^2}\right\} \\
&= -\frac{1}{t}\mathscr{L}^{-1}\left\{\frac{6s}{(s^2+1)(s^2+4)}\right\} \\
&= -\frac{1}{t}\mathscr{L}^{-1}\left\{\frac{2s}{s^2+1}-\frac{2s}{s^2+4}\right\} \\
&= -\frac{2}{t}\mathscr{L}^{-1}\left\{\frac{s}{s^2+1}\right\} + \frac{2}{t}\mathscr{L}^{-1}\left\{\frac{s}{s^2+4}\right\} \\
&= -\frac{2}{t}\cos t + \frac{2}{t}\cos 2t \\
&= \frac{2}{t}(\cos 2t - \cos t).
\end{aligned}$$

66. Applying Theorem 7.7 and letting $g(t) = 1$ and $f(t) = te^{-t}$ in Theorem 7.9 we obtain

$$\begin{aligned}
\mathscr{L}\left\{t\int_0^t \tau e^{-\tau}\,d\tau\right\} &= -\frac{d}{ds}\mathscr{L}\left\{\int_0^t \tau e^{-\tau}\,d\tau\right\} \\
&= -\frac{d}{ds}\left[\frac{\mathscr{L}\left\{te^{-t}\right\}}{s}\right] \\
&= -\frac{d}{ds}\left[\frac{1}{s(s+1)^2}\right] \\
&= \frac{3s+1}{s^2(s+1)^3}.
\end{aligned}$$

72. By the convolution theorem it follows that

$$\mathscr{L}\left\{e^{2t} * \sin t\right\} = \mathscr{L}\left\{e^{2t}\right\} \cdot \mathscr{L}\{\sin t\} = \frac{1}{s-2} \cdot \frac{1}{s^2+1} = \frac{1}{(s-2)(s^2+1)}.$$

78. Let

$$F(s) = G(s) = \frac{1}{(s+1)}.$$

Then $\mathscr{L}^{-1}\{F(s)\} = \mathscr{L}^{-1}\{G(s)\} = f(t) = g(t) = e^{-t}$. By the convolution theorem we have

$$\mathscr{L}^{-1}\left\{\frac{1}{(s+1)^2}\right\} = \int_0^t f(\tau)g(t-\tau)\,d\tau = \int_0^t e^{-\tau}e^{-(t-\tau)}\,dt = e^{-t}\int_0^t e^{-\tau+\tau}\,d\tau = e^{-t}\int_0^t d\tau = te^{-t}.$$

84. We identify the periodic function $f(t)$ with period 2 as

$$f(t) = \begin{cases} t, & 0 \le t \le 1 \\ -t+2, & 1 \le t \le 2 \end{cases}$$

and $f(t+2) = f(t)$. Then

$$\begin{aligned}
\mathscr{L}\{f\} &= \frac{1}{1-e^{-2s}}\int_0^2 e^{-st}f(t)\,dt \\
&= \frac{1}{1-e^{-2s}}\left[\int_0^1 te^{-st}\,dt + \int_1^2 (2-t)e^{-st}\,dt\right] \\
&= \frac{1}{1-e^{-2s}}\left[-\frac{t}{s}e^{-st} - \frac{1}{s^2}e^{-st}\right]_0^1 + \frac{1}{1-e^{-2s}}\left[-\frac{2}{s}e^{-st} + \frac{t}{s}e^{-st} + \frac{1}{s^2}e^{-st}\right]_1^2 \\
&= \frac{1}{1-e^{-2s}}\left[-\frac{1}{s}e^{-s} - \frac{1}{s^2}e^{-s} + \frac{1}{s^2}\right] + \frac{1}{1-e^{-2s}}\left[\frac{1}{s}e^{-s} + \frac{1}{s^2}e^{-2s} - \frac{1}{s^2}e^{-s}\right] \\
&= \frac{1}{1-e^{-2s}}\left[\frac{1}{s^2} - \frac{2}{s^2}e^{-s} + \frac{1}{s^2}e^{-2s}\right] \\
&= \frac{(1-e^{-s})^2}{s^2(1-e^{-2s})}.
\end{aligned}$$

Exercises 7.4

2. Taking the Laplace transform of both sides of the differential equation, we obtain

$$\begin{aligned}\mathscr{L}\left\{\frac{dy}{dt}\right\} + 2\mathscr{L}\{y\} &= \mathscr{L}\{t\}\\ sY(s) - y(0) + 2Y(s) &= \frac{1}{s^2}\\ (s+2)Y(s) &= -1 + \frac{1}{s^2}\\ Y(s) &= \frac{1-s^2}{s^2(s+2)}.\end{aligned}$$

By partial fractions

$$\frac{1-s^2}{s^2(s+2)} = \frac{A}{s} + \frac{B}{s^2} + \frac{C}{s+2}$$

and

$$1 - s^2 = As(s+2) + B(s+2) + Cs^2.$$

Setting $s = 0$ and $s = -2$ gives, in turn, $B = 1/2$ and $C = -3/4$. By equating coefficients of s^2 we get $-1 = A + C$, and so we find $A = -1/4$. Therefore

$$\begin{aligned}y(t) &= -\frac{1}{4}\mathscr{L}^{-1}\left\{\frac{1}{s}\right\} + \frac{1}{2}\mathscr{L}^{-1}\left\{\frac{1}{s^2}\right\} - \frac{3}{4}\mathscr{L}^{-1}\left\{\frac{1}{s+2}\right\}\\ &= -\frac{1}{4} + \frac{1}{2}t - \frac{3}{4}e^{-2t}.\end{aligned}$$

8. Taking the Laplace transform of both sides of the differential equation, we obtain

$$\begin{aligned}\mathscr{L}\{y''\} - \mathscr{L}\{4y'\} + \mathscr{L}\{4y\} &= \mathscr{L}\left\{t^3\right\}\\ s^2Y(s) - sy(0) - y'(0) - 4sY(s) + 4y(0) + 4Y(s) &= \frac{6}{s^4}\\ (s^2 - 4s + 4)Y(s) &= s - 4 + \frac{6}{s^4}\\ (s-2)^2Y(s) &= \frac{s^5 - 4s^4 + 6}{s^4}\\ Y(s) &= \frac{s^5 - 4s^4 + 6}{s^4(s-2)^2}.\end{aligned}$$

Now, by partial fractions

$$\frac{s^5 - 4s^4 + 6}{s^4(s-2)^2} = \frac{A}{s} + \frac{B}{s^2} + \frac{C}{s^3} + \frac{D}{s^4} + \frac{E}{s-2} + \frac{F}{(s-2)^2}$$

and

$$s^5 - 4s^4 + 6 = As^3(s-2)^2 + Bs^2(s-2)^2 + Cs(s-2)^2$$
$$+ D(s-2)^2 + Es^4(s-2) + Fs^4.$$

Setting $s = 0$ and $s = 2$ gives $D = 3/2$ and $F = -13/8$, respectively. Equating coefficients of s, s^2, s^3, and s^5 gives

$$0 = 4C - 4D$$
$$0 = 4B - 4C + D$$
$$0 = 4A - 4B + C$$
$$1 = A + E.$$

Using the known values of D and F these equations yield $C = 3/2$, $B = 9/8$, $A = 3/4$, and $E = 1/4$. Hence,

$$y(t) = \frac{3}{4}\mathscr{L}^{-1}\left\{\frac{1}{s}\right\} + \frac{9}{8}\mathscr{L}^{-1}\left\{\frac{1}{s^2}\right\} + \frac{3}{4}\mathscr{L}^{-1}\left\{\frac{2!}{s^3}\right\}$$
$$+ \frac{1}{4}\mathscr{L}^{-1}\left\{\frac{3!}{s^4}\right\} + \frac{1}{4}\mathscr{L}^{-1}\left\{\frac{1}{s-2}\right\} - \frac{13}{8}\mathscr{L}^{-1}\left\{\frac{1}{(s-2)^2}\right\}$$
$$= \frac{3}{4} + \frac{9}{8}t + \frac{3}{4}t^2 + \frac{1}{4}t^3 + \frac{1}{4}e^{2t} - \frac{13}{8}te^{2t}.$$

16. Taking the Laplace transform of both sides of the differential equation we obtain

$$\mathscr{L}\{y'''\} + \mathscr{L}\{2y''\} - \mathscr{L}\{y'\} - 2\mathscr{L}\{y\} = \mathscr{L}\{\sin 3t\}$$
$$s^3Y(s) - s^2y(0) - sy'(0) - y''(0) + 2s^2Y(s)$$
$$-2sy(0) - 2y'(0) - sY(s) + y(0) - 2Y(s) = \frac{3}{s^2+9}$$
$$(s^3 + 2s^2 - s - 2)Y(s) - 1 = \frac{3}{s^2+9}$$
$$(s-1)(s+1)(s+2)Y(s) = 1 + \frac{3}{s^2+9}$$
$$Y(s) = \frac{1}{(s-1)(s+1)(s+2)}$$
$$+ \frac{3}{(s-1)(s+1)(s+2)(s^2+9)}.$$

By partial fractions we find

$$Y(s) = \frac{1/6}{s-1} - \frac{1/2}{s+1} + \frac{1/3}{s+2} + \frac{1/20}{s-1} - \frac{3/20}{s+2} + \frac{1/13}{s+2} + \frac{3s/130 - 3/65}{s^2+9}$$
$$= \frac{13/60}{s-1} - \frac{13/20}{s+1} + \frac{16/39}{s+2} + \frac{3s/130}{s^2+9} - \frac{3/65}{s^2+9}.$$

Then

$$y(t) = \mathscr{L}^{-1}\{Y(s)\}$$
$$= \frac{13}{60}e^{t} - \frac{13}{20}e^{-t} + \frac{16}{39}e^{-2t} + \frac{3}{130}\cos 3t - \frac{1}{65}\sin 3t.$$

24. Taking the Laplace transform of both sides of the differential equation, we obtain

$$\mathscr{L}\{y''\} - \mathscr{L}\{5y'\} + \mathscr{L}\{6y\} = \mathscr{L}\{\mathscr{U}(t-1)\}$$
$$s^2Y(s) - sy(0) - y'(0) - 5sY(s) + 5y(0) + 6Y(s) = \frac{e^{-s}}{s}$$
$$(s^2 - 5s + 6)Y(s) = 1 + \frac{e^{-s}}{s}$$
$$Y(s) = \frac{1}{(s-2)(s-3)} + \frac{e^{-s}}{s(s-2)(s-3)}$$
$$= \frac{-1}{s-2} + \frac{1}{s-3} + \left[\frac{1/6}{s} - \frac{1/2}{s-2} + \frac{1/3}{s-3}\right]e^{-s}$$

Therefore

$$y(t) = -\mathscr{L}^{-1}\left\{\frac{1}{s-2}\right\} + \mathscr{L}^{-1}\left\{\frac{1}{s-3}\right\} + \frac{1}{6}\mathscr{L}^{-1}\left\{\frac{e^{-s}}{s}\right\} - \frac{1}{2}\mathscr{L}^{-1}\left\{\frac{e^{-s}}{s-2}\right\} + \frac{1}{3}\mathscr{L}^{-1}\left\{\frac{e^{-s}}{s-3}\right\}$$
$$= -e^{2t} + e^{3t} + \frac{1}{6}\mathscr{U}(t-1) - \frac{1}{2}e^{2(t-1)}\mathscr{U}(t-1) + \frac{1}{3}e^{3(t-1)}\mathscr{U}(t-1)$$
$$= \begin{cases} -e^{2t} + e^{3t}, & 0 \le t < 1 \\ -e^{2t} + e^{3t} + \frac{1}{6} - \frac{1}{2}e^{2(t-1)} + \frac{1}{3}e^{3(t-1)}, & t \ge 1. \end{cases}$$

28. Taking the Laplace transform of both sides of the differential equation and letting $c = y'(0)$ we obtain

$$\mathscr{L}\{y''\} - \mathscr{L}\{9y'\} + \mathscr{L}\{20y\} = \mathscr{L}\{1\}$$
$$s^2Y(s) - sy(0) - y'(0) - 9sY(s) + 9y(0) + 20Y(s) = \frac{1}{s}$$
$$s^2Y(s) - c - 9sY(s) + 20Y(s) = \frac{1}{s}$$
$$(s^2 - 9s + 20)Y(s) = \frac{1}{s} + c$$
$$Y(s) = \frac{1}{s(s^2 - 9s + 20)} + \frac{c}{s^2 - 9s + 20}$$
$$= \frac{1}{s(s-4)(s-5)} + \frac{c}{(s-4)(s-5)}$$
$$= \frac{1/20}{s} - \frac{1/4}{s-4} + \frac{1/5}{s-5} - \frac{c}{s-4} + \frac{c}{s-5}.$$

Therefore

$$y(t) = \frac{1}{20}\mathscr{L}^{-1}\{s\} - \frac{1}{4}\mathscr{L}^{-1}\{s-4\} + \frac{1}{5}\mathscr{L}^{-1}\{s-5\} - c\mathscr{L}^{-1}\{s-4\} + c\mathscr{L}^{-1}\{s-5\}$$

$$= \frac{1}{20} - \frac{1}{4}e^{4t} + \frac{1}{5}e^{5t} - c\left(e^{4t} - e^{5t}\right).$$

To find c we compute

$$y'(t) = -e^{4t} + e^{5t} - c\left(4e^{4t} - 5e^{5t}\right)$$

and let $y'(1) = 0$. Then

$$0 = -e^4 + e^5 - c\left(4e^4 - 5e^5\right)$$

and

$$c = \frac{e^5 - e^4}{4e^4 - 5e^5} = \frac{e-1}{4-5e}.$$

Thus,

$$y(t) = \frac{1}{20} - \frac{1}{4}e^{4t} + \frac{1}{5}e^{5t} - \frac{e-1}{4-5e}\left(e^{4t} - e^{5t}\right)$$

$$= \frac{1}{20} + \frac{e}{4(4-5e)}e^{4t} - \frac{1}{5(4-5e)}e^{5t}.$$

32. Taking the Laplace transform of both sides of the equation we obtain

$$\mathscr{L}\{f(t)\} + 2\mathscr{L}\left\{\int_0^t f(\tau)\cos(t-\tau)\,d\tau\right\} = 4\mathscr{L}\left\{e^{-t}\right\} + \mathscr{L}\{\sin t\}$$

$$F(s) + 2F(s)\frac{s}{s^2+1} = \frac{4}{s+1} + \frac{1}{s^2+1}$$

$$\frac{s^2+2s+1}{s^2+1}F(s) = \frac{4}{s+1} + \frac{1}{s^2+1}$$

$$\frac{(s+1)^2}{s^2+1}F(s) = \frac{4}{s+1} + \frac{1}{s^2+1}$$

$$F(s) = \frac{4s^2+4}{(s+1)^3} + \frac{1}{(s+1)^2}$$

By partial fractions,

$$F(s) = \frac{4}{s+1} - \frac{8}{(s+1)^2} + \frac{8}{(s+1)^3} + \frac{1}{(s+1)^2}$$

$$= \frac{4}{s+1} - \frac{7}{(s+1)^2} + \frac{8}{(s+1)^3}.$$

Hence we find

$$f(t) = 4\mathscr{L}^{-1}\left\{\frac{1}{s+1}\right\} - 7\mathscr{L}^{-1}\left\{\frac{1}{(s+1)^2}\right\} + \frac{8}{2}\mathscr{L}^{-1}\left\{\frac{2!}{(s+1)^3}\right\}$$

$$= 4e^{-t} - 7te^{-t} + 4t^2e^{-t}.$$

36. Taking the Laplace transform of both sides of the equation, we obtain

$$\begin{aligned}
\mathscr{L}\{t\} - \mathscr{L}\{2f(t)\} &= \mathscr{L}\left\{\int_0^t \left(e^\tau - e^{-\tau}\right) f(t-\tau)\,d\tau\right\} \\
\frac{1}{s^2} - 2F(s) &= \mathscr{L}\left\{e^t - e^{-t}\right\} F(s) \\
\frac{1}{s^2} - 2F(s) &= \frac{2}{s^2-1}F(s) \\
\left[2 + \frac{2}{s^2-1}\right] F(s) &= \frac{1}{s^2} \\
\frac{2s^2}{s^2-1}F(s) &= \frac{1}{s^2} \\
F(s) &= \frac{s^2-1}{2s^4} \\
&= \frac{1/2}{s^2} - \frac{1/2}{s^4}.
\end{aligned}$$

Therefore

$$\begin{aligned}
f(t) &= \frac{1}{2}\mathscr{L}^{-1}\left\{\frac{1}{s^2}\right\} - \frac{1}{12}\mathscr{L}^{-1}\left\{\frac{3!}{s^4}\right\} \\
&= \frac{1}{2}t - \frac{1}{12}t^3.
\end{aligned}$$

40. From equation (3) in the text we have

$$0.005\frac{di}{dt} + i + \frac{1}{0.02}\int_0^t i(\tau)\,d\tau = 100[t - (t-1)\,\mathscr{U}(t-1)]$$

or

$$\frac{di}{dt} + 200i + 10{,}000\int_0^t i(\tau)\,d\tau = 20{,}000[t - (t-1)\,\mathscr{U}(t-1)].$$

Taking the Laplace transform of both sides of this equation we obtain

$$sI(s) + 200I(s) + 10{,}000\frac{I(s)}{s} = 20{,}000\left[\frac{1}{s^2} - e^{-s}\frac{1}{s^2}\right].$$

Then

$$\begin{aligned}
\left(s + 200 + \frac{10{,}000}{s}\right) I(s) &= \frac{20{,}000}{s^2}\left(1 - e^{-s}\right) \\
\left(s^2 + 200s + 10{,}000\right) I(s) &= \frac{20{,}000}{s}\left(1 - e^{-s}\right) \\
(s+100)^2 I(s) &= \frac{20{,}000}{s}\left(1 - e^{-s}\right) \\
I(s) &= \frac{20{,}000}{s(s+100)^2}\left(1 - e^{-s}\right).
\end{aligned}$$

Using partial fractions we obtain

$$I(s) = \left[\frac{2}{s} - \frac{2}{s+100} - \frac{200}{(s+100)^2}\right]\left(1 - e^{-s}\right).$$

Thus

$$i(t) = \mathscr{L}^{-1}\left\{\frac{2}{s} - \frac{2}{s+100} - \frac{200}{(s+100)^2} - \frac{2}{s}e^{-s} + \frac{2}{s+100}e^{-s} + \frac{200}{(s+100)^2}e^{-s}\right\}$$

$$= 2 - 2e^{-100t} - 200te^{-100t} - 2\,\mathscr{U}(t-1) + 2e^{-100(t-1)}\,\mathscr{U}(t-1)$$

$$+ 200(t-1)e^{-100(t-1)}\,\mathscr{U}(t-1).$$

44. The initial-value problem is

$$\frac{di}{dt} + \frac{R}{L}i = \frac{1}{L}E(t), \quad i(0) = 0.$$

Taking the Laplace transform of both sides of the equation, we obtain

$$sI(s) + \frac{R}{L}I(s) = \frac{1}{L}\mathscr{L}\{E(t)\}.$$

From Problem 81, Exercise 7.3, we have

$$\mathscr{L}\{E(t)\} = \frac{1 - e^{-s}}{s(1 + e^{-s})}.$$

Thus

$$\left(s + \frac{R}{L}\right)I(s) = \frac{1}{L}\cdot\frac{1 - e^{-s}}{s(1 + e^{-s})}$$

and

$$I(s) = \frac{1}{L}\cdot\frac{1 - e^{-s}}{s(s + R/L)(1 + e^{-s})}$$

$$= \frac{1}{L}\cdot\frac{1 - e^{-s}}{s(s + R/L)}\cdot\frac{1}{1 + e^{-s}}$$

$$= \frac{1}{L}\left[\frac{L/R}{s} - \frac{L/R}{s + R/L}\right](1 - e^{-s})(1 - e^{-s} + e^{-2s} - e^{-3s} + \cdots)$$

$$= \frac{1}{R}\left[\frac{1}{s} - \frac{1}{s + R/L}\right](1 - 2e^{-s} + 2e^{-2s} - 2e^{-3s} + \cdots).$$

Therefore

$$i(t) = \frac{1}{R}[1 - 2\,\mathscr{U}(t-1) + 2\,\mathscr{U}(t-2) - 2\,\mathscr{U}(t-3) + \cdots]$$

$$- \frac{1}{R}\Big[e^{-Rt/L} + 2e^{-R(t-1)/L}\,\mathscr{U}(t-1) - 2e^{-R(t-2)/L}\,\mathscr{U}(t-2)$$

$$+ 2e^{-R(t-3)/L}\,\mathscr{U}(t-3) + \cdots\Big]$$

$$= \frac{1}{R}\left(1 - e^{-Rt/L}\right) + \frac{2}{R}\sum_{n=1}^{\infty}(-1)^n\left(1 - e^{-R(t-n)/L}\right)\mathscr{U}(t-n).$$

50. Recall from Chapter 5 that

$$mx'' = -kx + f(t).$$

Now $m = W/g = 32/32 = 1$ slug, and $32 = 2k$ so that $k = 16$ lb/ft. Thus, the differential equation is

$$x'' + 16x = f(t).$$

The initial conditions are $x(0) = 0$, $x'(0) = 0$. Also, since

$$f(t) = \begin{cases} \sin t, & 0 \le t < 2\pi \\ 0, & t \ge 2\pi \end{cases}$$

and $\sin t = \sin(t - 2\pi)$ we can write

$$f(t) = \sin t - \sin(t - 2\pi)\,\mathscr{U}(t - 2\pi).$$

Therefore,

$$\mathscr{L}\{x''\} + \mathscr{L}\{16x\} = \mathscr{L}\{\sin t\} - \mathscr{L}\{\sin(t-2\pi)\,\mathscr{U}(t-2\pi)\}$$

$$s^2X(s) + 16X(s) = \frac{1}{s^2+1} - \frac{1}{s^2+1}e^{-2\pi s}$$

$$X(s) = \frac{1}{(s^2+16)(s^2+1)} - \frac{1}{(s^2+16)(s^2+1)}e^{-2\pi s}$$

$$= \frac{-1/15}{s^2+16} + \frac{1/15}{s^2+1} - \left[\frac{-1/15}{s^2+16} + \frac{1/15}{s^2+1}\right]e^{-2\pi s},$$

and

$$x(t) = -\frac{1}{60}\mathscr{L}^{-1}\left\{\frac{4}{s^2+16}\right\} + \frac{1}{15}\mathscr{L}^{-1}\left\{\frac{1}{s^2+1}\right\} + \frac{1}{60}\mathscr{L}^{-1}\left\{\frac{4e^{-2\pi s}}{s^2+16}\right\} - \frac{1}{15}\mathscr{L}^{-1}\left\{\frac{e^{-2\pi s}}{s^2+1}\right\}$$

$$= -\frac{1}{60}\sin 4t + \frac{1}{15}\sin t + \frac{1}{60}\sin 4(t-2\pi)\,\mathscr{U}(t-2\pi) - \frac{1}{15}\sin(t-2\pi)\,\mathscr{U}(t-2\pi)$$

$$= \begin{cases} -\frac{1}{60}\sin 4t + \frac{1}{15}\sin t, & 0 \le t < 2\pi \\ 0, & t \ge 2\pi. \end{cases}$$

52. Recall from Chapter 5 that

$$mx'' = -kx + f(t).$$

Now $m = W/g = 16/32 = 1/2$ slug, and $k = 4.5$, so the differential equation is

$$\frac{1}{2}x'' + 4.5x = 4\sin 3t + 2\cos 3t$$

or

$$x'' + 9x = 8\sin 3t + 4\cos 3t.$$

Taking the Laplace transform of both sides of the equation and using the fact that $x(0) = x'(0) = 0$ we obtain

$$s^2X(s) + 9X(s) = \frac{24}{s^2+9} + \frac{4s}{s^2+9}.$$

Then

$$X(s) = \frac{4s+24}{(s^2+9)^2} = \frac{2}{3}\cdot\frac{2(3)s}{(s^2+9)^2} + \frac{12}{27}\cdot\frac{2(3)^3}{(s^2+9)^2}$$

and

$$x(t) = \frac{2}{3}t\sin 3t + \frac{4}{9}(\sin 3t - 3t\cos 3t)$$

$$= \frac{2}{3}t\sin 3t + \frac{4}{9}\sin 3t - \frac{4}{3}t\cos 3t.$$

56. The differential equation is

$$EI\frac{d^4y}{dx^4} = w_0 \qquad \text{or} \qquad \frac{d^4y}{dx^4} = \frac{w_0}{EI}.$$

Taking the Laplace transform of both sides and using $y(0) = y''(0) = 0$ we obtain

$$s^4\,\mathscr{L}\{y\} - s^2y'(0) - y'''(0) = \frac{w_0}{EI}\frac{1}{s}.$$

Letting $y'(0) = c_1$ and $y'''(0) = c_2$ we have

$$\mathscr{L}\{y\} = \frac{w_0}{EI}\frac{1}{s^5} + \frac{c_1}{s^2} + \frac{c_2}{s^4}.$$

Then

$$y(x) = \frac{w_0}{EI}\frac{1}{4!}t^4 + c_1t + \frac{c_2}{3!}t^3$$

$$= \frac{w_0}{24EI}t^4 + c_1t + \frac{c_2}{6}t^3.$$

To find c_1 and c_2 we compute

$$y''(x) = \frac{w_0}{2EI}t^2 + c_2t.$$

Then $y(L) = y''(L) = 0$ yields the system

$$c_1L + c_2\frac{L^3}{6} = -\frac{w_0L^4}{24EI}$$

$$c_2L = -\frac{w_0L^2}{2EI}.$$

Solving for c_1 and c_2 we obtain $c_1 = w_0L^3/24EI$ and $c_2 = -w_0L/2EI$. Thus

$$y(x) = \frac{w_0}{24EI}x^4 - \frac{w_0 L}{12EI}x^3 + \frac{w_0 L^3}{24EI}x.$$

Exercises 7.5

6. Taking the Laplace transform of both sides of the differential equation, we obtain

$$s^2Y(s) - sy(0) - y'(0) + Y(s) = e^{-2\pi s} + e^{-4\pi s}$$

$$\left(s^2+1\right)Y(s) - s = e^{-2\pi s} + e^{-4\pi s}$$

$$Y(s) = \frac{s}{s^2+1} + \frac{e^{-2\pi s}}{s^2+1} + \frac{e^{-4\pi s}}{s^2+1}.$$

Then

$$y(t) = \mathscr{L}^{-1}\left\{\frac{s}{s^2+1}\right\} + \mathscr{L}^{-1}\left\{\frac{e^{-2\pi s}}{s^2+1}\right\} + \mathscr{L}^{-1}\left\{\frac{e^{-4\pi s}}{s^2+1}\right\}$$

$$= \cos t + \sin(t-2\pi)\,\mathscr{U}(t-2\pi) + \sin(t-4\pi)\,\mathscr{U}(t-4\pi).$$

8. Taking the Laplace transform of both sides of the differential equation, we obtain

$$s^2Y(s) - sy(0) - y'(0) - 2sY(s) + y(0) = \frac{1}{s} + e^{-2s}$$

$$\left(s^2-2s\right)Y(s) - 1 = \frac{1}{s} + e^{-2s}$$

$$Y(s) = \frac{1}{s^2-2s} + \frac{1}{s\left(s^2-2s\right)} + \frac{e^{-2s}}{s^2-2s}$$

$$= \frac{1}{s(s-2)} + \frac{1}{s^2(s-2)} + \frac{e^{-2s}}{s(s-2)}$$

$$= \frac{1/2}{s-2} - \frac{1/2}{s} + \frac{1/4}{s-2} - \frac{1/2}{s^2} - \frac{1/4}{s} + \frac{e^{-2s}/2}{s-2} - \frac{e^{-2s}/2}{s}$$

$$= \frac{3/4}{s-2} - \frac{3/4}{s} - \frac{1/2}{s^2} + \frac{e^{-2s}/2}{s-2} - \frac{e^{-2s}/2}{s}.$$

Then

$$y(t) = \frac{3}{4}\mathscr{L}^{-1}\left\{\frac{1}{s-2}\right\} - \frac{3}{4}\mathscr{L}^{-1}\left\{\frac{1}{s}\right\} - \frac{1}{2}\mathscr{L}^{-1}\left\{\frac{1}{s^2}\right\} + \frac{1}{2}\mathscr{L}^{-1}\left\{\frac{e^{-2s}}{s-2}\right\} - \frac{1}{2}\mathscr{L}^{-1}\left\{\frac{e^{-2s}}{s}\right\}$$

$$= \frac{3}{4}e^{2t} - \frac{3}{4} - \frac{1}{2}t + \frac{1}{2}e^{2(t-2)}\,\mathscr{U}(t-2) - \frac{1}{2}\mathscr{U}(t-2).$$

14. Taking the Laplace transform of both sides of the differential equation, we obtain

$$y(x) = \frac{1}{2}c_1x^2 + \frac{1}{6}c_2x^3 + \frac{P_0}{EI}\frac{1}{6}\left(x - \frac{L}{2}\right)^3 \mathscr{U}\left(x - \frac{L}{2}\right).$$

To find c_1 and c_2 we compute

$$y'(x) = c_1x + \frac{1}{2}c_2x^2 + \frac{P_0}{EI}\frac{1}{2}\left(x - \frac{L}{2}\right)^2, \quad \text{for } x \geq \frac{L}{2}.$$

Then $y(L) = y'(L) = 0$ yields the system

$$\frac{1}{2}c_1L^2 + \frac{1}{6}c_2L^3 + \frac{P_0}{EI}\frac{1}{6}\frac{L^3}{8} = 0$$

$$c_1L + \frac{1}{2}c_2L^2 + \frac{P_0}{EI}\frac{1}{2}\frac{L^3}{4} = 0.$$

Solving this system gives $c_1 = P_0L/8EI$ and $c_2 = -P_0/2EI$. Hence

$$y(x) = \frac{P_0L}{EI}\frac{1}{16}x^2 - \frac{P_0}{EI}\frac{1}{12}x^3 + \frac{P_0}{EI}\frac{1}{6}\left(x - \frac{L}{2}\right)^3 \mathscr{U}\left(x - \frac{L}{2}\right)$$

$$= \begin{cases} \frac{P_0}{EI}\left(\frac{L}{16}x^2 - \frac{1}{12}x^3\right), & 0 \leq x < L/2 \\ \frac{P_0}{EI}\left(\frac{L}{16}x^2 - \frac{1}{12}x^3\right) + \frac{P_0}{6EI}\left(x - \frac{L}{2}\right)^3, & L/2 \leq x \leq L. \end{cases}$$

18. Taking the Laplace transform of both sides of the differential equation, we obtain

$$s^2Y(s) + \omega^2Y(s) = 1$$

$$Y(s) = \frac{1}{s^2 + \omega^2}.$$

Then

$$y(t) = \mathscr{L}^{-1}\left\{\frac{1}{s^2 + \omega^2}\right\} = \frac{1}{\omega}\sin\omega t$$

and

$$y'(t) = \cos\omega t.$$

Thus, $y'(0) = \cos 0 = 1 \neq 0$.

8 Systems of Linear Differential Equations

Exercises 8.1

4. The given system is

$$Dx - 4y = 1$$

$$x + Dy = 2.$$

Operating on the second equation by D and subtracting the result from the first equation eliminates x from the system:

$$-D^2y - 4y = 1 \qquad \text{or} \qquad D^2y + 4y = -1.$$

For the latter equation $y_c = c_1 \cos 2t + c_2 \sin 2t$. Furthermore, the method of undetermined coefficients yields a constant solution $y_p = -1/4$. Hence

$$y(t) = c_1 \cos 2t + c_2 \sin 2t - \frac{1}{4}.$$

Now the second equation of the system provides a means for determining x: $x = 2 - Dy$. After differentiating y we find

$$x(t) = 2c_1 \sin 2t - 2c_2 \cos 2t + 2.$$

A solution of the system is then

$$x(t) = 2c_1 \sin 2t - 2c_2 \cos 2t + 2$$

$$y(t) = c_1 \cos 2t + c_2 \sin 2t - \frac{1}{4}.$$

8. To solve the system

$$\begin{aligned} \left(D^2 + 5\right) x + Dy &= 0 \\ (D + 1)x + (D - 4)y &= 0 \end{aligned} \tag{1}$$

we operate on the first equation by $D - 4$ and the second by D:

$$\left(D^3 - 4D^2 + 5D - 20\right) x + (D - 4)Dy = 0$$

$$\left(D^2 + D\right) x + D(D - 4)y = 0.$$

Subtracting the second equation from the first we have

$$\left(D^3 - 5D^2 + 4D - 20\right) x = 0$$

or

$$(D-5)\left(D^2+4\right) x = 0.$$

The roots of the characteristic equation are 5, $2i$, and $-2i$. Thus,

$$x(t) = c_1 e^{5t} + c_2 \cos 2t + c_3 \sin 2t.$$

To find y we operate on the first equation in (1) by $D+1$ and the second by D^2+5:

$$\begin{aligned} (D+1)(D^2+5)x + (D^2+D)y &= 0 \\ (D^2+5)(D+1)x + (D^3-4D^2+5D-20)y &= 0. \end{aligned} \tag{2}$$

Subtracting the first equation in (2) from the second we have

$$\left(D^3 - 5D^2 + 4D - 20\right) y = 0$$

or

$$(D-5)\left(D^2+4\right) y = 0.$$

The roots of the characteristic equation are 5, $2i$, and $-2i$. Thus,

$$y(t) = c_4 e^{5t} + c_5 \cos 2t + c_6 \sin 2t.$$

Substituting x and y into the second equation in (1) we have

$$\begin{aligned} &5c_1e^{5t} - 2c_2 \sin 2t + 2c_3 \cos 2t + c_1 e^{5t} + c_2 \cos 2t + c_3 \sin 2t \\ &\qquad + 5c_4 e^{5t} - 2c_5 \sin 2t + 2c_6 \cos 2t - 4c_4 e^{5t} - 4c_5 \cos 2t - 4c_6 \sin 2t = 0 \end{aligned}$$

or

$$(6c_1 + c_4)e^{5t} + (c_2 + 2c_3 - 4c_5 + 2c_6)\cos 2t + (-2c_2 + c_3 - 2c_5 - 4c_6)\sin 2t = 0.$$

Hence

$$\begin{aligned} 6c_1 + c_4 &= 0 \\ c_2 + 2c_3 - 4c_5 + 2c_6 &= 0 \\ -2c_2 + c_3 - 2c_5 - 4c_6 &= 0. \end{aligned} \tag{3}$$

We want to solve for c_4, c_5, and c_6 in terms of c_1, c_2, and c_3. From the first equation is (3) we have

$$c_4 = -6c_1.$$

We write the second and third equations in (3) as

$$4c_5 - 2c_6 = c_2 + 2c_3$$
$$2c_5 + 4c_6 = -2c_2 + c_3$$

and solve for c_5 and c_6. We obtain

$$c_5 - \frac{1}{2}c_3 \qquad \text{and} \qquad c_6 = -\frac{1}{2}c_2.$$

Therefore, the solution of the system (1) is

$$x(t) = c_1e^{5t} + c_2\cos 2t + c_3\sin 2t$$
$$y(t) = -6c_1e^{5t} + \frac{1}{2}c_3\cos 2t - \frac{1}{2}c_2\sin 2t.$$

20. We write the system in the form

$$\begin{aligned} (D+1)x \qquad\qquad\qquad - z &= 0 \\ (D+1)y \quad - z &= 0 \\ x \qquad\quad - y \; + Dz &= 0. \end{aligned}$$

By Cramer's rule we can write

$$\begin{vmatrix} D+1 & 0 & -1 \\ 0 & D+1 & -1 \\ 1 & -1 & D \end{vmatrix} x = \begin{vmatrix} 0 & 0 & -1 \\ 0 & D+1 & -1 \\ 0 & -1 & D \end{vmatrix}.$$

The second determinant is 0 since its first column is all zeros. Expanding the first determinant by the first row, we have

$$\left[(D+1)\left(D^2+D-1\right)-(-D-1)\right]x = 0$$
$$\left[(D+1)\left(D^2+D-1\right)+(D+1)\right]x = 0$$
$$(D+1)\left(D^2+D\right)x = 0$$
$$D(D+1)^2x = 0.$$

The roots of the characteristic equation are 0, -1 and -1. Thus,

$$x = c_1 + c_2e^{-t} + c_3te^{-t}.$$

To find z we use $(D+1)x - z = 0$:

$$\begin{aligned} z &= x' + x \\ &= -c_2e^{-t} - c_3te^{-t} + c_3e^{-t} + c_1 + c_2e^{-t} + c_3te^{-t} \\ &= c_1 + c_3e^{-t}. \end{aligned}$$

To find y we use $x - y + Dz = 0$:

$$\begin{aligned} y &= x + Dz \\ &= c_1 + c_2e^{-t} + c_3te^{-t} - c_3e^{-t} \\ &= c_1 + (c_2 - c_3)e^{-t} + c_3te^{-t}. \end{aligned}$$

Therefore, the solution of the system is

$$\begin{aligned} x(t) &= c_1 + c_2e^{-t} + c_3te^{-t} \\ y(t) &= c_1 + (c_2 - c_3)e^{-t} + c_3te^{-t} \\ z(t) &= c_1 + c_3e^{-t}. \end{aligned}$$

22. To solve the system

$$\begin{aligned} Dx \qquad\quad &- 2Dy = t^2 \\ (D+1)x \quad &- 2(D+1)y = 1 \end{aligned}$$

we operate on the first equation by $D+1$ and on the second equation by D:

$$\begin{aligned} (D+1)Dx - 2(D+1)Dy &= (D+1)t^2 = 2t + t^2 \\ D(D+1)x - 2D(D+1)y &= D(1) = 0. \end{aligned}$$

Subtracting the second equation from the first, we have

$$0 = 2t + t^2.$$

Therefore, the system has no solution.

Exercises 8.2

2. Letting $X(s) = \mathscr{L}\{x(t)\}$ and $Y(s) = \mathscr{L}\{y(t)\}$ we transform the system

$$\begin{aligned} \frac{dx}{dt} &= 2y + e^t \\ \frac{dy}{dt} &= 8x - t \\ x(0) = 1,\ & y(0) = 1 \end{aligned} \tag{1}$$

into

$$\begin{aligned} sX(s) - x(0) &= 2Y(s) + \frac{1}{s-1} \\ sY(s) - y(0) &= 8X(s) - \frac{1}{s^2} \end{aligned}$$

or

$$\begin{aligned} sX(s) - 2Y(s) &= 1 + \frac{1}{s-1} \\ -8X(s) + sY(s) &= 1 - \frac{1}{s^2}. \end{aligned}$$

We multiply the first equation by 8 and the second equation by s:

$$\begin{aligned} 8sX(s) - 16Y(s) &= 8 + \frac{8}{s-1} \\ -8sX(s) + s^2Y(s) &= s - \frac{1}{s}. \end{aligned}$$

Adding, we obtain

$$\left(s^2 - 16\right) Y(s) = s + 8 + \frac{8}{s-1} - \frac{1}{s}$$

or

$$Y(s) = \frac{s+8}{s^2-16} + \frac{8}{(s-1)\left(s^2-16\right)} - \frac{1}{s\left(s^2-16\right)}.$$

By partial fractions

$$\frac{8}{(s-1)\left(s^2-16\right)} = \frac{-8/15}{s-1} + \frac{1/3}{s-4} + \frac{1/5}{s+4}$$

and

$$\frac{1}{s\left(s^2-16\right)} = \frac{-1/16}{s} + \frac{1/32}{s-4} + \frac{1/32}{s+4}.$$

Thus,

$$\begin{aligned} Y(s) &= \frac{s+8}{s^2-16} - \frac{8/15}{s-1} + \frac{1/3}{s-4} + \frac{1/5}{s+4} + \frac{1/16}{s} - \frac{1/32}{s-4} - \frac{1/32}{s+4} \\ &= \frac{s+8}{s^2-16} + \frac{1/16}{s} - \frac{8/15}{s-1} + \frac{29/96}{s-4} + \frac{27/160}{s+4}, \end{aligned}$$

and so

$$y(t) = \mathscr{L}^{-1}\left\{\frac{s+8}{s^2-16}\right\} + \frac{1}{16}\mathscr{L}^{-1}\left\{\frac{1}{s}\right\} - \frac{8}{15}\mathscr{L}^{-1}\left\{\frac{1}{s-1}\right\}$$
$$+ \frac{29}{96}\mathscr{L}^{-1}\left\{\frac{1}{s-4}\right\} + \frac{27}{160}\mathscr{L}^{-1}\left\{\frac{1}{s+4}\right\}$$
$$= \cosh 4t + 2\sinh 4t + \frac{1}{16} - \frac{8}{15}e^t + \frac{29}{96}e^{4t} + \frac{27}{160}e^{-4t}$$
$$= \frac{1}{2}e^{4t} + \frac{1}{2}e^{-4t} + e^{4t} - e^{-4t} + \frac{1}{16} - \frac{8}{15}e^t + \frac{29}{96}e^{4t} + \frac{27}{160}e^{-4t}$$
$$= \frac{1}{16} - \frac{8}{15}e^t + \frac{173}{96}e^{4t} - \frac{53}{160}e^{-4t}.$$

To find x we compute $y'(t)$ and use the second equation in (1):

$$y'(t) = -\frac{8}{15}e^t + \frac{173}{24}e^{4t} + \frac{53}{40}e^{-4t} = 8x - t$$
$$x(t) = \frac{1}{8}t - \frac{1}{15}e^t + \frac{173}{192}e^{4t} + \frac{53}{320}e^{-4t}.$$

Therefore, the solution of the system is

$$x(t) = \frac{1}{8}t - \frac{1}{15}e^t + \frac{173}{192}e^{4t} + \frac{53}{320}e^{-4t}$$
$$y(t) = \frac{1}{16} - \frac{8}{15}e^t + \frac{173}{96}e^{4t} - \frac{53}{160}e^{-4t}.$$

6. Letting $X(s) = \mathscr{L}\{x(t)\}$ and $Y(s) = \mathscr{L}\{y(t)\}$ we transform the system

$$\begin{aligned} \frac{dx}{dt} + x - \frac{dy}{dt} + y &= 0 \\ \frac{dx}{dt} + \frac{dy}{dt} + 2y &= 0 \\ x(0) = 0,\ y(0) &= 1 \end{aligned} \tag{2}$$

into

$$sX(s) - x(0) + X(s) - sY(s) + y(0) + Y(s) = 0$$
$$sX(s) - x(0) + sY(s) - y(0) + 2Y(s) = 0$$

or

$$(s+1)X(s) - (s-1)Y(s) = -1$$
$$sX(s) + (s+2)Y(s) = 1.$$

We multiply the first equation by s and the second equation by $s+1$:

$$\begin{aligned} s(s+1)X(s) - \left(s^2 - s\right)Y(s) &= -s \\ (s+1)sX(s) + \left(s^2 + 3s + 2\right)Y(s) &= s+1. \end{aligned} \tag{3}$$

Subtracting the first equation from the second, we obtain

$$\left(2s^2 + 2s + 2\right)Y(s) = 2s + 1$$

so that

$$Y(s) = \frac{s + 1/2}{s^2 + s + 1} = \frac{s + 1/2}{(s + 1/2)^2 + 3/4}.$$

Thus

$$y(t) = \mathscr{L}^{-1}\left\{\frac{s + 1/2}{(s + 1/2)^2 + 3/4}\right\} = e^{-t/2}\cos\frac{\sqrt{3}}{2}t.$$

To obtain $x(t)$ we multiply the first equation in (3) by $s+2$ and the second equation by $s-1$:

$$\begin{aligned} \left(s^2 + 3s + 2\right)X(s) - (s+2)(s-1)Y(s) &= -s - 2 \\ \left(s^2 - s\right)X(s) + (s-1)(s+2)Y(s) &= s - 1. \end{aligned}$$

Adding we obtain

$$\left(2s^2 + 2s + 2\right)X(s) = -3,$$

so that

$$X(s) = \frac{-3/2}{s^2 + s + 1} = \frac{-3/2}{(s + 1/2)^2 + 3/4}.$$

Thus,

$$\begin{aligned} x(t) &= \mathscr{L}^{-1}\left\{\frac{-3/2}{(s + 1/2)^2 + 3/4}\right\} \\ &= -\sqrt{3}\,\mathscr{L}^{-1}\left\{\frac{\sqrt{3}/2}{(s + 1/2)^2 + 3/4}\right\} \\ &= -\sqrt{3}\,e^{-t/2}\sin\frac{\sqrt{3}}{2}t. \end{aligned}$$

Therefore, the solution of the system (2) is

$$\begin{aligned} x(t) &= -\sqrt{3}\,e^{-t/2}\sin\frac{\sqrt{3}}{2}t \\ y(t) &= e^{-t/2}\cos\frac{\sqrt{3}}{2}t. \end{aligned}$$

12. Letting $X(s) = \mathscr{L}\{x(t)\}$ and $Y(s) = \mathscr{L}\{y(t)\}$ we transform the system

$$\begin{aligned} \frac{dx}{dt} &= 4x - 2y + 2\mathcal{U}(t-1) \\ \frac{dy}{dt} &= 3x - y + \mathcal{U}(t-1) \\ x(0) &= 0,\ y(0) = \frac{1}{2} \end{aligned} \tag{4}$$

into

$$sX(s) - x(0) = 4X(s) - 2Y(s) + \frac{2}{s}e^{-s}$$

$$sY(s) - y(0) = 3X(s) - Y(s) + \frac{1}{s}e^{-s}$$

or

$$\begin{aligned} (s-4)X(s) + 2Y(s) &= \frac{2}{s}e^{-s} \\ -3X(s) + (s+1)Y(s) &= \frac{1}{2} + \frac{1}{s}e^{-s}. \end{aligned} \tag{5}$$

Multiplying the first equation by 3 and the second equation by $s-4$ we obtain

$$3(s-4)X(s) + 6Y(s) = \frac{6}{s}e^{-s}$$

$$-3(s-4)X(s) + \left(s^2 - 3s - 4\right)Y(s) = \frac{s-4}{2} + \frac{s-4}{s}e^{-s}.$$

Adding, we have

$$\left(s^2 - 3s + 2\right)Y(s) = \frac{s-4}{2} + \frac{s+2}{s}e^{-s}$$

or

$$(s-1)(s-2)Y(s) = \frac{s-4}{2} + \frac{s+2}{s}e^{-s}$$

Then,

$$Y(s) = \frac{1}{2}\frac{s-4}{(s-1)(s-2)} + \frac{s+2}{s(s-1)(s-2)}e^{-s}.$$

By partial fractions

$$\frac{s-4}{(s-1)(s-2)} = \frac{3}{s-1} + \frac{-2}{s-2}$$

and

$$\frac{s+2}{s(s-1)(s-2)} = \frac{1}{s} + \frac{-3}{s-1} + \frac{2}{s-2}.$$

Thus,

$$Y(s) = \frac{3}{2}\frac{1}{s-1} - \frac{1}{s-2} + \left(\frac{1}{s} - \frac{3}{s-1} + \frac{2}{s-2}\right)e^{-s}$$

and so

$$y(t) = \frac{3}{2}\mathscr{L}^{-1}\left\{\frac{1}{s-1}\right\} - \mathscr{L}^{-1}\left\{\frac{1}{s-2}\right\} + \mathscr{L}^{-1}\left\{\left(\frac{1}{s} - \frac{3}{s-1} + \frac{2}{s-2}\right)e^{-s}\right\}$$
$$= \frac{3}{2}e^t - e^{2t} + \left(1 - 3e^{t-1} + 2e^{2(t-1)}\right)\mathscr{U}(t-1).$$

To find $x(t)$ we multiply the first equation in (5) by $s+1$ and the second equation by 2:

$$\left(s^2 - 3s - 4\right)X(s) + 2(s+1)Y(s) = \frac{2(s+1)}{s}e^{-s}$$
$$-6X(s) + 2(s+1)Y(s) = 1 + \frac{2}{3}e^{-s}.$$

Subtracting the second equation from the first, we obtain

$$\left(s^2 - 3s + 2\right)X(s) = -1 + 2e^{-s}.$$

Then,

$$X(s) = \frac{-1}{(s-1)(s-2)} + \frac{2}{(s-1)(s-2)}e^{-s}$$
$$= \frac{1}{s-1} - \frac{1}{s-2} + \left(\frac{-2}{s-1} + \frac{2}{s-2}\right)e^{-s}$$

and so

$$x(t) = \mathscr{L}^{-1}\left\{\frac{1}{s-1}\right\} - \mathscr{L}^{-1}\left\{\frac{1}{s-2}\right\} + \mathscr{L}^{-1}\left\{\left(-\frac{2}{s-1} + \frac{2}{s-2}\right)e^{-s}\right\}$$
$$= e^t - e^{2t} + \left(-2e^{t-1} + 2e^{2(t-1)}\right)\mathscr{U}(t-1).$$

Therefore, the solution of (4) is

$$x(t) = e^t - e^{2t} - 2\left(e^{t-1} - e^{2t-2}\right)\mathscr{U}(t-1)$$
$$y(t) = \frac{3}{2}e^t - e^{2t} + \left(1 - 3e^{t-1} + 2e^{2t-2}\right)\mathscr{U}(t-1).$$

16. (a) From Kirchoff's law we have

$$L\frac{di_1}{dt} + R_1 i_2 = E$$
$$L\frac{di_1}{dt} + R_2 i_3 + \frac{1}{C}q = E.$$

Subtracting the two equations and differentiating the resulting equation gives

$$-R_1\frac{di_2}{dt} + R_2\frac{di_3}{dt} + \frac{1}{C}i_3 = 0.$$

Since $i_1 = i_2 + i_3$ the first equation becomes

$$L\frac{di_2}{dt} + L\frac{di_3}{dt} + R_1 i_2 = E.$$

Hence, the system is

$$L\frac{di_2}{dt} + L\frac{di_3}{dt} + R_1 i_2 = E$$

$$-R_1\frac{di_2}{dt} + R_2\frac{di_3}{dt} + \frac{1}{C}i_3 = 0.$$

(b) Transforming the system

$$\frac{di_2}{dt} + 10i_2 + \frac{di_3}{dt} = 120 - 120\,\mathcal{U}(t-2)$$

$$-10\frac{di_2}{dt} + 5\frac{di_3}{dt} + 5i_3 = 0$$

yields

$$(s+10)I_2(s) + \qquad sI_3(s) = \frac{120}{s} - \frac{120}{s}e^{-2s}$$

$$-10I_2(s) + (5s+5)I_3(s) = 0.$$

Solving for $I_2(s)$ and using partial fractions gives

$$I_2(s) = \frac{12}{s} + \frac{144}{3s+5} - \frac{60}{s+2} - \left[\frac{12}{s} + \frac{144}{3s+5} - \frac{60}{s+2}\right]e^{-2s}$$

and so

$$i_2(t) = 12 + 48e^{-5t/3} - 60e^{-2t} - \left[12 + 48e^{-5(t-2)/3} - 60e^{-2t(t-2)}\right]\mathcal{U}(t-2).$$

Similarly we obtain

$$I_3(s) = \frac{720}{3s+5} - \frac{240}{s+2} - \left[\frac{720}{3s+5} - \frac{240}{s+2}\right]e^{-2s}$$

and consequently

$$i_3(t) = 240e^{-5t/3} - 240e^{-2t} - \left[240e^{-5(t-2)/3} - 240e^{-2(t-2)}\right]\mathcal{U}(t-2).$$

(c) Using $i_1 = i_2 + i_3$ and the results in (a) and (b) it follows that

$$i_1(t) = 12 + 288e^{-5t/3} - 300e^{-2t} - \left[12 + 288e^{-5(t-2)/3} - 300e^{-2(t-2)}\right]\mathcal{U}(t-2).$$

18. Summing the voltage drops across the smaller loop gives

$$L\frac{di_1}{dt} + Ri_2 = E(t). \qquad (6)$$

Summing across the larger loop gives

$$L\frac{di_1}{dt} + \frac{q}{C} = E(t). \tag{7}$$

Substituting equation (6) into equation (7) gives

$$L\frac{di_1}{dt} + \frac{q}{C} = L\frac{di_1}{dt} + Ri_2$$

or

$$q = RCi_2.$$

Differentiating we have

$$\frac{dq}{dt} = RC\frac{di_2}{dt}.$$

Now, $dq/dt = i_3$ and $i_1 = i_2 + i_3$ so

$$i_3 = RC\frac{di_2}{dt}$$

$$i_1 - i_2 = RC\frac{di_2}{dt}.$$

Therefore, the system of equations is

$$L\frac{di_1}{dt} + Ri_2 = E(t)$$

$$RC\frac{di_2}{dt} + i_2 - i_1 = 0.$$

Exercises 8.3

8. Let $y = x_1$ and $y' = x_2$. Then $y' = x_1' = x_2$ and $y'' = x_2'$ so that

$$t^2y'' + ty' + \left(t^2 - 4\right)y = 0 \tag{1}$$

becomes

$$t^2x_2' + tx_2 + \left(t^2 - 4\right)x_1 = 0.$$

We see then that (1) can be reduced to the system

$$x_1' = x_2$$

$$x_2' = -\frac{t_2 - 4}{t^2}x_1 - \frac{1}{t}x_2.$$

12. We write the system as

$$\begin{aligned} D^2x - 2D^2y &= \sin t \\ D^2x + D^2y &= \cos t. \end{aligned} \tag{2}$$

Subtracting the first equation from the second, we have

$$3D^2y = \cos t - \sin t$$

or

$$D^2y = \frac{1}{3}\cos t - \frac{1}{3}\sin t.$$

Adding the first equation to twice the second equation in (2) we have

$$3D^2x = 2\cos t + \sin t$$

or

$$D^2x = \frac{2}{3}\cos t + \frac{1}{3}\sin t.$$

If we let

$$Dx = u \qquad \text{and} \qquad Dy = v$$

then

$$Du = D^2x = \frac{2}{3}\cos t + \frac{1}{3}\sin t$$

and

$$Dv = D^2y = \frac{1}{3}\cos t - \frac{1}{3}\sin t.$$

The system (2) can be written in the normal form

$$\begin{aligned} Dx &= u \\ Dy &= v \\ Du &= \frac{2}{3}\cos t + \frac{1}{3}\sin t \\ Dv &= \frac{1}{3}\cos t - \frac{1}{3}\sin t. \end{aligned}$$

Exercises 8.4

6. (a) $\mathbf{AB} = (5 \quad -6 \quad 7)\begin{pmatrix} 3 \\ 4 \\ -1 \end{pmatrix} = (5\cdot 3 - 6\cdot 4 - 7\cdot 1) = (15 - 24 - 7) = (-16)$

(b) $\mathbf{BA} = \begin{pmatrix} 3 \\ 4 \\ -1 \end{pmatrix}(5 \quad -6 \quad 7) = \begin{pmatrix} 3\cdot 5 & 3\cdot(-6) & 3\cdot 7 \\ 4\cdot 5 & 4\cdot(-6) & 4\cdot 7 \\ (-1)\cdot 5 & (-1)\cdot(-6) & (-1)\cdot 7 \end{pmatrix} = \begin{pmatrix} 15 & -18 & 21 \\ 20 & -24 & 28 \\ -5 & 6 & -7 \end{pmatrix}$

(c) $(\mathbf{BA})\mathbf{C} = \begin{pmatrix} 15 & -18 & 21 \\ 20 & -24 & 28 \\ -5 & 6 & -7 \end{pmatrix}\begin{pmatrix} 1 & 2 & 4 \\ 0 & 1 & -1 \\ 3 & 2 & 1 \end{pmatrix} = \begin{pmatrix} 78 & 54 & 99 \\ 104 & 72 & 132 \\ -26 & -18 & -33 \end{pmatrix}$

(d) Since $\mathbf{AB}$ is a 1×1 matrix and $\mathbf{C}$ is a 3×3 matrix, the product $(\mathbf{AB})\mathbf{C}$ is not defined.

8. (a) $\mathbf{A} + \mathbf{B}^T = \begin{pmatrix} 1 & 2 \\ 2 & 4 \end{pmatrix} + \begin{pmatrix} -2 & 5 \\ 3 & 7 \end{pmatrix} = \begin{pmatrix} -1 & 7 \\ 5 & 11 \end{pmatrix}$

(b) $2\mathbf{A}^T - \mathbf{B}^T = 2\begin{pmatrix} 1 & 2 \\ 2 & 4 \end{pmatrix} - \begin{pmatrix} -2 & 5 \\ 3 & 7 \end{pmatrix} = \begin{pmatrix} 2 & 4 \\ 4 & 8 \end{pmatrix} - \begin{pmatrix} -2 & 5 \\ 3 & 7 \end{pmatrix} = \begin{pmatrix} 4 & -1 \\ 1 & 1 \end{pmatrix}$

(c) $\mathbf{A}^T(\mathbf{A} - \mathbf{B}) = \begin{pmatrix} 1 & 2 \\ 2 & 4 \end{pmatrix}\left[\begin{pmatrix} 1 & 2 \\ 2 & 4 \end{pmatrix} - \begin{pmatrix} -2 & 3 \\ 5 & 7 \end{pmatrix}\right] = \begin{pmatrix} 1 & 2 \\ 2 & 4 \end{pmatrix}\begin{pmatrix} 3 & -1 \\ -3 & -3 \end{pmatrix} = \begin{pmatrix} -3 & -7 \\ -6 & -14 \end{pmatrix}$

12. $3t\begin{pmatrix} 2 \\ t \\ -1 \end{pmatrix} + (t-1)\begin{pmatrix} -1 \\ -t \\ 3 \end{pmatrix} - 2\begin{pmatrix} 3t \\ 4 \\ -5t \end{pmatrix} = \begin{pmatrix} 6t \\ 3t^2 \\ -3t \end{pmatrix} + \begin{pmatrix} -t+1 \\ -t^2+t \\ 3t-3 \end{pmatrix} - \begin{pmatrix} 6t \\ 8 \\ -10t \end{pmatrix} = \begin{pmatrix} -t+1 \\ 2t^2+t-8 \\ 10t-3 \end{pmatrix}$

20. Since $\det \mathbf{A} = 27 \neq 0$, $\mathbf{A}$ is non-singular. To find $\mathbf{A}^{-1}$ we first compute the co-factors.

$$\mathbf{A}_{11} = \begin{vmatrix} 1 & 0 \\ 5 & -1 \end{vmatrix} = -1 \qquad \mathbf{A}_{12} = -\begin{vmatrix} 4 & 0 \\ -2 & -1 \end{vmatrix} = 4 \qquad \mathbf{A}_{13} = \begin{vmatrix} 4 & 1 \\ -2 & 5 \end{vmatrix} = 22$$

$$\mathbf{A}_{21} = -\begin{vmatrix} 2 & 1 \\ 5 & -1 \end{vmatrix} = 7 \qquad \mathbf{A}_{22} = \begin{vmatrix} 3 & 1 \\ -2 & -1 \end{vmatrix} = -1 \qquad \mathbf{A}_{23} = -\begin{vmatrix} 3 & 2 \\ -2 & 5 \end{vmatrix} = -19$$

$$\mathbf{A}_{31} = \begin{vmatrix} 2 & 1 \\ 1 & 0 \end{vmatrix} = -1 \qquad \mathbf{A}_{32} = -\begin{vmatrix} 3 & 1 \\ 4 & 0 \end{vmatrix} = 4 \qquad \mathbf{A}_{33} = \begin{vmatrix} 3 & 2 \\ 4 & 1 \end{vmatrix} = -5$$

Then

$$\mathbf{A}^{-1} = \frac{1}{\det \mathbf{A}} \begin{pmatrix} -1 & 4 & 22 \\ 7 & -1 & -19 \\ -1 & 4 & -5 \end{pmatrix}^T = \frac{1}{27} \begin{pmatrix} -1 & 7 & -1 \\ 4 & -1 & 4 \\ 22 & -19 & -5 \end{pmatrix}.$$

24. Since

$$\begin{aligned} \det \mathbf{A}(t) &= 2e^{2t} \sin^2 t + 2e^{2t} \cos^2 t \\ &= 2e^{2t} \left(\sin^2 t + \cos^2 t \right) \\ &= 2e^{2t} > 0 \end{aligned}$$

for all t, $\mathbf{A}(t)$ is non-singular. From (3) in the text we have

$$(\mathbf{A}(t))^{-1} = \frac{1}{2e^{2t}} \begin{pmatrix} e^t \sin t & 2e^t \cos t \\ -e^t \cos t & 2e^t \sin t \end{pmatrix} = \frac{1}{2} \begin{pmatrix} e^{-t} \sin t & 2e^{-t} \cos t \\ -e^{-t} \cos t & 2e^{-t} \sin t \end{pmatrix}.$$

28. $\dfrac{d}{dt} \begin{pmatrix} 5te^{2t} \\ t \sin 3t \end{pmatrix} = \begin{pmatrix} 10te^{2t} + 5e^{2t} \\ 3t \cos 3t + \sin 3t \end{pmatrix}.$

36. We indicate the row-operations on a matrix using the notation on page 389 in the text.

$$\left(\begin{array}{ccc|c} 1 & 0 & 2 & 8 \\ 1 & 2 & -2 & 4 \\ 2 & 5 & -6 & 6 \end{array}\right) \overset{\substack{A_{12}(-1) \\ A_{13}(-2)}}{\Longrightarrow} \left(\begin{array}{ccc|c} 1 & 0 & 2 & 8 \\ 0 & 2 & -4 & -4 \\ 0 & 5 & -10 & -10 \end{array}\right) \overset{\substack{M_2(\frac{1}{2}) \\ M_3(\frac{1}{5})}}{\Longrightarrow} \left(\begin{array}{ccc|c} 1 & 0 & 2 & 8 \\ 0 & 1 & -2 & -2 \\ 0 & 1 & -2 & -2 \end{array}\right)$$

$$\overset{A_{23}(-1)}{\Longrightarrow} \left(\begin{array}{ccc|c} 1 & 0 & 2 & 8 \\ 0 & 1 & -2 & -2 \\ 0 & 0 & 0 & 0 \end{array}\right)$$

By letting $z = t$, the last matrix indicates that a solution to the system is $x = 8 - 2t$, $y = -2 + 2t$, $z = t$.

40. We row-reduce the augmented matrix.

$$\left(\begin{array}{cccc|c} 1 & 1 & -1 & 3 & 1 \\ 0 & 1 & -1 & -4 & 0 \\ 1 & 2 & -2 & -1 & 6 \\ 4 & 7 & -7 & 0 & 9 \end{array}\right) \overset{\substack{A_{13}(-1) \\ A_{14}(-4)}}{\Longrightarrow} \left(\begin{array}{cccc|c} 1 & 1 & -1 & 3 & 1 \\ 0 & 1 & -1 & -4 & 0 \\ 0 & 1 & -1 & -4 & 5 \\ 0 & 3 & -3 & -12 & 5 \end{array}\right) \overset{A_{23}(-1)}{\Longrightarrow} \left(\begin{array}{cccc|c} 1 & 1 & -1 & 3 & 1 \\ 0 & 1 & -1 & -4 & 0 \\ 0 & 0 & 0 & 0 & 5 \\ 0 & 3 & -3 & -12 & 5 \end{array}\right)$$

Inspection of the third row of the last matrix shows that the system has no solution.

42. We first compute the determinant of $\mathbf{A} - \lambda\mathbf{I}$.

$$\det(\mathbf{A} - \lambda\mathbf{I}) = \begin{vmatrix} 2-\lambda & 1 \\ 2 & 1-\lambda \end{vmatrix} = \lambda^2 - 3\lambda = \lambda(\lambda - 3).$$

Thus, the eigenvalues are $\lambda_1 = 0$ and $\lambda_2 = 3$. For $\lambda_1 = 0$ we must solve

$$2k_1 + k_2 = 0$$
$$2k_1 + k_2 = 0.$$

From this system we see $k_2 = -2k_1$. Choosing $k_1 = 1$ then gives $k_2 = -2$. Hence

$$\mathbf{K}_1 = \begin{pmatrix} 1 \\ -2 \end{pmatrix}.$$

For $\lambda_2 = 3$ we have

$$-k_1 + k_2 = 0$$
$$2k_1 - 2k_2 = 0,$$

and so we see $k_1 = k_2$. The choice $k_2 = 1$ implies $k_1 = 1$. Therefore,

$$\mathbf{K}_2 = \begin{pmatrix} 1 \\ 1 \end{pmatrix}.$$

50. We first compute the determinant of $\mathbf{A} - \lambda\mathbf{I}$.

$$\det(\mathbf{A} - \lambda\mathbf{I}) = \begin{vmatrix} 2-\lambda & -1 & 0 \\ 5 & 2-\lambda & 4 \\ 0 & 1 & 2-\lambda \end{vmatrix} = (2-\lambda)(\lambda^2 - 4\lambda + 5)$$

Solving $(2-\lambda)(\lambda^2 - 4\lambda + 5) = 0$ we find the eigenvalues $\lambda_1 = 2$, $\lambda_2 = 2 + i$, $\lambda_3 = 2 - i$. For $\lambda_1 = 2$ the system

$$0k_1 - k_2 + 0k_3 = 0$$
$$5k_1 + 0k_2 + 4k_3 = 0$$
$$0k_1 + k_2 + 0k_3 = 0$$

yields $k_2 = 0$ and $k_1 = -(4/5)k_3$. With $k_3 = 5$ we get

$$\mathbf{K}_1 = \begin{pmatrix} -4 \\ 0 \\ 5 \end{pmatrix}.$$

For $\lambda_2 = 2 + i$ the system

$$-ik_1 - k_2 + 0k_3 = 0$$
$$5k_1 - ik_2 + 4k_3 = 0$$
$$0k_1 + k_2 - ik_3 = 0$$

implies $k_1 = -(1/i)k_2$ and $k_3 = (1/i)k_2$. Choosing $k_2 = i$ we obtain

$$\mathbf{K}_2 = \begin{pmatrix} -1 \\ i \\ 1 \end{pmatrix}.$$

Finally, for $\lambda_3 = 2 - i$ we see from

$$ik_1 - k_2 + 0k_3 = 0$$
$$5k_1 + ik_2 + 4k_3 = 0$$
$$0k_1 + k_2 + ik_3 = 0$$

that $k_1 = (1/i)k_2$ and $k_3 = -(1/i)k_2$. The choice $k_2 = -i$ gives

$$\mathbf{K}_3 = \begin{pmatrix} -1 \\ -i \\ 1 \end{pmatrix}.$$

Exercises 8.5

6. The system in matrix form is

$$\frac{d}{dt}\begin{pmatrix} x \\ y \end{pmatrix} = \begin{pmatrix} -3 & 4 \\ 5 & 9 \end{pmatrix}\begin{pmatrix} x \\ y \end{pmatrix} + \begin{pmatrix} e^{-t}\sin 2t \\ 4e^{-t}\cos 2t \end{pmatrix}.$$

10. The system is

$$\frac{dx}{dt} = 3x - 7y + 4\sin t + (t-4)e^{4t}$$
$$\frac{dy}{dt} = x + y + 8\sin t + (2t+1)e^{4t}.$$

18. We compute the Wronskian

$$W(\mathbf{X}_1, \mathbf{X}_2) = \begin{vmatrix} e^t & 2e^t + 8te^t \\ -e^t & 6e^t - 8te^t \end{vmatrix}$$
$$= 6e^{2t} - 8te^{2t} + 2e^{2t} + 8te^{2t}$$
$$= 8e^{2t}.$$

Since $8e^{2t} > 0$ for all t, $\mathbf{X}_1$ and $\mathbf{X}_2$ are linearly independent and form a fundamental set of solutions.

28. The fundamental matrix and its inverse are

$$\mathbf{\Phi}(t) = \begin{pmatrix} -e^{-t} & e^{5t} \\ e^{-t} & e^{5t} \end{pmatrix}$$

and

$$\mathbf{\Phi}^{-1}(t) = \frac{1}{-2e^{4t}} \begin{pmatrix} e^{5t} & -e^{5t} \\ -e^{-t} & -e^{-t} \end{pmatrix} = \frac{1}{2} \begin{pmatrix} -e^{t} & e^{t} \\ e^{-5t} & e^{-5t} \end{pmatrix}.$$

32. From Problem 28 we know that the general solution of the system is

$$\mathbf{X}(t) = c_1 \begin{pmatrix} -1 \\ 1 \end{pmatrix} e^{-t} + c_2 \begin{pmatrix} 1 \\ 1 \end{pmatrix} e^{5t}.$$

When $t = 0$ and $\mathbf{X}(0) = \begin{pmatrix} 1 \\ 0 \end{pmatrix}$ we have

$$c_1 \begin{pmatrix} -1 \\ 1 \end{pmatrix} + c_2 \begin{pmatrix} 1 \\ 1 \end{pmatrix} = \begin{pmatrix} 1 \\ 0 \end{pmatrix}$$

or

$$-c_1 + c_2 = 1$$

$$c_1 + c_2 = 0.$$

The solution of this system is $c_1 = -1/2$ and $c_2 = 1/2$. Thus,

$$\mathbf{V}_1 = -\frac{1}{2} \begin{pmatrix} -1 \\ 1 \end{pmatrix} e^{-t} + \frac{1}{2} \begin{pmatrix} 1 \\ 1 \end{pmatrix} e^{5t} = \begin{pmatrix} \frac{1}{2}e^{-t} + \frac{1}{2}e^{5t} \\ -\frac{1}{2}e^{-t} + \frac{1}{2}e^{5t} \end{pmatrix}.$$

When $\mathbf{X}(0) = \begin{pmatrix} 0 \\ 1 \end{pmatrix}$ we have

$$c_1 \begin{pmatrix} -1 \\ 1 \end{pmatrix} + c_2 \begin{pmatrix} 1 \\ 1 \end{pmatrix} = \begin{pmatrix} 0 \\ 1 \end{pmatrix}$$

or

$$-c_1 + c_2 = 0$$

$$c_1 + c_2 = 1.$$

The solution of this system is $c_1 = c_2 = 1/2$. Thus

$$\mathbf{V}_2 = \frac{1}{2} \begin{pmatrix} -1 \\ 1 \end{pmatrix} e^{-t} + \frac{1}{2} \begin{pmatrix} 1 \\ 1 \end{pmatrix} e^{5t} = \begin{pmatrix} -\frac{1}{2}e^{-t} + \frac{1}{2}e^{5t} \\ \frac{1}{2}e^{-t} + \frac{1}{2}e^{5t} \end{pmatrix}.$$

Therefore

$$\mathbf{\Psi}(t) = \frac{1}{2} \begin{pmatrix} e^{-t} + e^{5t} & -e^{-t} + e^{5t} \\ -e^{-t} + e^{5t} & e^{-t} + e^{5t} \end{pmatrix}.$$

Exercises 8.6

6. The characteristic equation is

$$\begin{vmatrix} -6-\lambda & 2 \\ -3 & 1-\lambda \end{vmatrix} = 0$$

or

$$\lambda^2 + 5\lambda = 0.$$

The eigenvalues are $\lambda_1 = 0$ and $\lambda_2 = -5$. For $\lambda_1 = 0$ we have the system

$$-6k_1 + 2k_2 = 0$$
$$-3k_1 + k_2 = 0.$$

Choosing $k_1 = 1$ gives $k_2 = 3$. Thus, an eigenvector corresponding to $\lambda_1 = 0$ is

$$\mathbf{K}_1 = \begin{pmatrix} 1 \\ 3 \end{pmatrix}.$$

For $\lambda_2 = -5$ we have the system

$$-k_1 + 2k_2 = 0$$
$$-3k_1 + 6k_2 = 0.$$

Choosing $k_2 = 1$ gives $k_1 = 2$. Thus, an eigenvector corresponding to $\lambda_2 = -5$ is

$$\mathbf{K}_2 = \begin{pmatrix} 2 \\ 1 \end{pmatrix}.$$

The general solution of the system is

$$\begin{aligned} \mathbf{X}(t) &= c_1 \begin{pmatrix} 1 \\ 3 \end{pmatrix} e^{0t} + c_2 \begin{pmatrix} 2 \\ 1 \end{pmatrix} e^{-5t} \\ &= c_1 \begin{pmatrix} 1 \\ 3 \end{pmatrix} + c_2 \begin{pmatrix} 2 \\ 1 \end{pmatrix} e^{-5t}. \end{aligned}$$

10. The characteristic equation is

$$\begin{vmatrix} 1-\lambda & 0 & 1 \\ 0 & 1-\lambda & 0 \\ 1 & 0 & 1-\lambda \end{vmatrix} = 0$$

or

$$-\lambda^3 + 3\lambda^2 - 2\lambda = 0.$$

Factoring we have $-\lambda(\lambda-1)(\lambda-2)=0$. The eigenvalues are $\lambda_1=0$, $\lambda_2=1$, and $\lambda_3=2$. For $\lambda_1=0$ we have

$$\begin{aligned} k_1 \quad + k_3 &= 0 \\ k_2 \quad &= 0 \\ k_1 \quad + k_3 &= 0. \end{aligned}$$

Thus, $k_2=0$ and choosing $k_1=1$ we have $k_3=-1$. Hence, an eigenvector corresponding to $\lambda_1=0$ is

$$\mathbf{K}_1=\begin{pmatrix}1\\0\\-1\end{pmatrix}.$$

For $\lambda_2=1$ we have

$$\begin{aligned} k_3 &= 0 \\ 0 &= 0 \\ k_1 &= 0. \end{aligned}$$

Thus, $k_1=k_3=0$ and we may choose k_2 to be any real number. An eigenvector corresponding to $\lambda_2=1$ is

$$\mathbf{K}_2=\begin{pmatrix}0\\1\\0\end{pmatrix}.$$

For $\lambda_3=2$ we have

$$\begin{aligned} -k_1 \quad + k_3 &= 0 \\ -k_2 \quad &= 0 \\ k_1 \quad - k_3 &= 0. \end{aligned}$$

Thus, $k_2=0$ and choosing $k_1=1$ we have $k_3=1$. Hence, an eigenvector corresponding to $\lambda_3=2$ is

$$\mathbf{K}_3=\begin{pmatrix}1\\0\\1\end{pmatrix}.$$

The general solution of the system is

$$\mathbf{X}(t)=c_1\begin{pmatrix}1\\0\\-1\end{pmatrix}+c_2\begin{pmatrix}0\\1\\0\end{pmatrix}e^t+c_3\begin{pmatrix}1\\0\\1\end{pmatrix}e^{2t}.$$

14. The characteristic equation is

$$\begin{vmatrix}1-\lambda & 1 & 4\\ 0 & 2-\lambda & 0\\ 1 & 1 & 1-\lambda\end{vmatrix}=0$$

or

$$(\lambda - 2)(\lambda - 3)(\lambda + 1) = 0.$$

Corresponding to the eigenvalues $\lambda_1 = 2$, $\lambda_2 = 3$, and $\lambda_3 = -1$ we find, in turn, the eigenvectors

$$\mathbf{K}_1 = \begin{pmatrix} 5 \\ -3 \\ 2 \end{pmatrix}, \quad \mathbf{K}_2 = \begin{pmatrix} 2 \\ 0 \\ 1 \end{pmatrix}, \quad \text{and} \quad \mathbf{K}_3 = \begin{pmatrix} -2 \\ 0 \\ 1 \end{pmatrix}.$$

Hence the general solution of the system is

$$\mathbf{X}(t) = c_1 \begin{pmatrix} 5 \\ -3 \\ 2 \end{pmatrix} e^{2t} + c_2 \begin{pmatrix} 2 \\ 0 \\ 1 \end{pmatrix} e^{3t} + c_3 \begin{pmatrix} -2 \\ 0 \\ 1 \end{pmatrix} e^{-t}.$$

Now the initial condition

$$\mathbf{X}(0) = \begin{pmatrix} 1 \\ 3 \\ 0 \end{pmatrix}$$

implies

$$\begin{pmatrix} 1 \\ 3 \\ 0 \end{pmatrix} = c_1 \begin{pmatrix} 5 \\ -3 \\ 2 \end{pmatrix} + c_2 \begin{pmatrix} 2 \\ 0 \\ 1 \end{pmatrix} + c_3 \begin{pmatrix} -2 \\ 0 \\ 1 \end{pmatrix}$$

or

$$1 = 5c_1 + 2c_2 - 2c_3$$

$$3 = -3c_1$$

$$0 = 2c_1 + c_2 + c_3.$$

The solution to this algebraic system is $c_1 = -1$, $c_2 = 5/2$, $c_3 = -1/2$. Consequently, a solution of the initial-value problem is

$$\mathbf{X}(t) = -\begin{pmatrix} 5 \\ -3 \\ 2 \end{pmatrix} e^{2t} + \frac{5}{2} \begin{pmatrix} 2 \\ 0 \\ 1 \end{pmatrix} e^{3t} - \frac{1}{2} \begin{pmatrix} -2 \\ 0 \\ 1 \end{pmatrix} e^{-t}.$$

22. The characteristic equation is

$$\begin{vmatrix} 2-\lambda & 1 & 2 \\ 3 & -\lambda & 6 \\ -4 & 0 & -3-\lambda \end{vmatrix} = 0$$

or

$$(\lambda + 3)\left(\lambda^2 - 2\lambda + 5\right) = 0.$$

The eigenvalues are $\lambda_1 = -3$, $\lambda_2 = 1 + 2i$, $\lambda_3 = 1 - 2i$. For $\lambda_1 = -3$ we have

$$\begin{aligned} 5k_1 + k_2 + 2k_3 &= 0 \\ 3k_1 + 3k_2 + 6k_3 &= 0 \\ -4k_1 &= 0. \end{aligned}$$

Thus, $k_1 = 0$ and choosing $k_3 = 1$ we have $k_2 = -2$. Hence, an eigenvector corresponding to $\lambda_1 = -3$ is

$$\mathbf{K}_1 = \begin{pmatrix} 0 \\ -2 \\ 1 \end{pmatrix}.$$

For $\lambda_2 = 1 + 2i$ we have

$$\begin{aligned} (1 - 2i)k_1 + \qquad k_2 + \qquad 2k_3 &= 0 \\ 3k_1 - (1 + 2i)k_2 + \qquad 6k_3 &= 0 \\ -4k_1 \qquad\qquad - (4 + 2i)k_3 &= 0. \end{aligned}$$

Adding $1 + 2i$ times the first equation to the second equation, we have the system

$$\begin{aligned} (1 - 2i)k_1 + k_2 + \qquad 2k_3 &= 0 \\ 8k_1 \qquad + (8 + 4i)k_3 &= 0 \\ -4k_1 \qquad - (4 + 2i)k_3 &= 0. \end{aligned}$$

Since the second and third equations are now equivalent we may write the system in the form

$$\begin{aligned} (1 - 2i)k_1 + k_2 + \qquad 2k_3 &= 0 \\ 2k_1 \qquad + (2 + i)k_3 &= 0. \end{aligned}$$

Choosing $k_3 = 2$ we then have $k_1 = -2 - i$ and $k_2 = -(1 - 2i)k_1 - 2k_3 = -(1 - 2i)(-2 - i) - 6 = -2 - 3i$. Hence, an eigenvector corresponding to $\lambda_2 = 1 + 2i$ is

$$\mathbf{K}_2 = \begin{pmatrix} -2 - i \\ -2 - 3i \\ 2 \end{pmatrix}$$

and an eigenvector corresponding to $\lambda_3 = 1 - 2i$ is

$$\mathbf{K}_3 = \begin{pmatrix} -2 + i \\ -2 + 3i \\ 2 \end{pmatrix}.$$

Now

$$\mathbf{B}_1 = \frac{1}{2}\left[\begin{pmatrix} -2-i \\ -2-3i \\ 2 \end{pmatrix} + \begin{pmatrix} -2+i \\ -2+3i \\ 2 \end{pmatrix}\right] = \begin{pmatrix} -2 \\ -2 \\ 2 \end{pmatrix}$$

and

$$\mathbf{B}_2 = \frac{i}{2}\left[\begin{pmatrix} -2-i \\ -2-3i \\ 2 \end{pmatrix} - \begin{pmatrix} -2+i \\ -2+3i \\ 2 \end{pmatrix}\right] = \begin{pmatrix} 1 \\ 3 \\ 0 \end{pmatrix}.$$

The general solution is

$$\mathbf{X}(t) = c_1 \begin{pmatrix} 0 \\ -2 \\ 1 \end{pmatrix} e^{-3t} + c_2 \left[\begin{pmatrix} -2 \\ -2 \\ 2 \end{pmatrix} \cos 2t + \begin{pmatrix} 1 \\ 3 \\ 0 \end{pmatrix} \sin 2t\right] e^t + c_3 \left[\begin{pmatrix} 1 \\ 3 \\ 0 \end{pmatrix} \cos 2t - \begin{pmatrix} -2 \\ -2 \\ 2 \end{pmatrix} \sin 2t\right] e^t$$

$$= c_1 \begin{pmatrix} 0 \\ -2 \\ 1 \end{pmatrix} e^{-3t} + c_2 \begin{pmatrix} -2\cos 2t + \sin 2t \\ -2\cos 2t + 3\sin 2t \\ 2\cos 2t \end{pmatrix} c^t + c_3 \begin{pmatrix} \cos 2t + 2\sin 2t \\ 3\cos 2t + 2\sin 2t \\ -2\sin 2t \end{pmatrix} e^t.$$

28. The characteristic equation is

$$\begin{vmatrix} 6-\lambda & -1 \\ 5 & 4-\lambda \end{vmatrix} = 0$$

or $\lambda^2 - 10\lambda + 29 = 0$. The eigenvalues are $\lambda_1 = 5 + 2i$ and $\lambda_2 = 5 - 2i$. For $\lambda = 5 + 2i$ we have

$$\begin{aligned} (1-2i)k_1 - \qquad\qquad k_2 &= 0 \\ 5k_1 + (-1-2i)k_2 &= 0. \end{aligned}$$

For $k_1 = 1$ we have $k_2 = 1 - 2i$. Thus

$$\mathbf{K}_1 = \begin{pmatrix} 1 \\ 1-2i \end{pmatrix},$$

so

$$\mathbf{B}_1 = \frac{1}{2}\left[\begin{pmatrix} 1 \\ 1-2i \end{pmatrix} + \begin{pmatrix} 1 \\ 1+2i \end{pmatrix}\right] = \begin{pmatrix} 1 \\ 1 \end{pmatrix}$$

and

$$\mathbf{B}_2 = \frac{i}{2}\left[\begin{pmatrix} 1 \\ 1-2i \end{pmatrix} - \begin{pmatrix} 1 \\ 1+2i \end{pmatrix}\right] = \begin{pmatrix} 0 \\ 2 \end{pmatrix}.$$

Thus

$$\mathbf{X}(t) = c_1 \left[\begin{pmatrix} 1 \\ 1 \end{pmatrix} \cos 2t + \begin{pmatrix} 0 \\ 2 \end{pmatrix} \sin 2t\right] e^t + c_2 \left[\begin{pmatrix} 0 \\ 2 \end{pmatrix} \cos 2t - \begin{pmatrix} 1 \\ 1 \end{pmatrix} \sin 2t\right] e^t$$

$$= c_1 \begin{pmatrix} \cos 2t \\ \cos 2t + 2\sin 2t \end{pmatrix} e^{5t} + c_2 \begin{pmatrix} -\sin 2t \\ 2\cos 2t - \sin 2t \end{pmatrix} e^{5t}.$$

Now

$$\begin{pmatrix} -2 \\ 8 \end{pmatrix} = \mathbf{X}(0) = c_1 \begin{pmatrix} \cos 0 \\ \cos 0 + 2\sin 0 \end{pmatrix} e^0 + c_2 \begin{pmatrix} -\sin 0 \\ 2\cos 0 - \sin 0 \end{pmatrix} e^0$$

or

$$c_1 \begin{pmatrix} 1 \\ 1 \end{pmatrix} + c_2 \begin{pmatrix} 0 \\ 2 \end{pmatrix} = \begin{pmatrix} -2 \\ 8 \end{pmatrix}.$$

Solving the system

$$\begin{aligned} c_1 \qquad &= -2 \\ c_1 + 2c_2 &= 8 \end{aligned}$$

we obtain $c_1 = -2$ and $c_2 = 5$. Therefore, the solution of the initial value problem is

$$\mathbf{X}(t) = -2 \begin{pmatrix} \cos 2t \\ \cos 2t + 2\sin 2t \end{pmatrix} e^{5t} + 5 \begin{pmatrix} -\sin 2t \\ 2\cos 2t - \sin 2t \end{pmatrix} e^{5t}.$$

34. The characteristic equation is

$$\begin{vmatrix} 3-\lambda & 2 & 4 \\ 2 & -\lambda & 2 \\ 4 & 2 & 3-\lambda \end{vmatrix} = 0$$

or $(\lambda+1)^2(\lambda-8)$. The eigenvalues are $\lambda_1 = 8$ and $\lambda_2 = \lambda_3 = -1$. For $\lambda_1 = 8$ we have

$$\begin{aligned} -5k_1 + 2k_2 + 4k_3 &= 0 \\ 2k_1 - 8k_2 + 2k_3 &= 0 \\ 4k_1 + 2k_2 - 5k_3 &= 0. \end{aligned}$$

If we multiply the second equation by 1/2 and rearrange the equations, we have

$$\begin{aligned} k_1 - 4k_2 + k_3 &= 0 \\ -5k_1 + 2k_2 + 4k_3 &= 0 \\ 4k_1 + 2k_2 - 5k_3 &= 0. \end{aligned}$$

Now, adding 5 times the first equation to the second equation and -4 times the first equation to the third equation, we have

$$\begin{aligned} k_1 - 4k_2 + k_3 &= 0 \\ -18k_2 + 9k_3 &= 0 \\ 18k_2 - 9k_3 &= 0. \end{aligned}$$

This is the equivalent to the system

$$\begin{aligned} k_1 - 4k_2 + k_3 &= 0 \\ 2k_2 - k_3 &= 0. \end{aligned}$$

Choosing $k_2 = 1$ we have $k_3 = 2$ and $k_1 = 2$. Thus,

$$\mathbf{K}_1 = \begin{pmatrix} 2 \\ 1 \\ 2 \end{pmatrix}$$

is an eigenvector corresponding to $\lambda_1 = 8$. For $\lambda_2 = -1$ we have

$$\begin{aligned} 4k_1 + 2k_2 + 4k_3 &= 0 \\ 2k_1 + \ k_2 + 2k_3 &= 0 \\ 4k_1 + 2k_2 + 4k_3 &= 0. \end{aligned}$$

Since the first and third equations are multiples of the second equation, the system is equivalent to the system consisting of the single equation

$$2k_1 + k_2 + 2k_3 = 0.$$

Choosing $k_1 = 0$ and $k_3 = 1$ we have $k_2 = -2$. Thus,

$$\mathbf{K}_2 = \begin{pmatrix} 0 \\ -2 \\ 1 \end{pmatrix}$$

is an eigenvector corresponding to $\lambda_2 = -1$. Now, choosing $k_3 = 0$ and $k_1 = 1$, we have $k_2 = -2$. Thus,

$$\mathbf{K}_3 = \begin{pmatrix} 1 \\ -2 \\ 0 \end{pmatrix}$$

is also an eigenvector corresponding to $\lambda_3 = -1$. Since $\mathbf{K}_2$ and $\mathbf{K}_3$ are linearly independent, the general solution is

$$\mathbf{X}(t) = c_1 \begin{pmatrix} 2 \\ 1 \\ 2 \end{pmatrix} e^{8t} + c_2 \begin{pmatrix} 0 \\ -2 \\ 1 \end{pmatrix} e^{-t} + c_3 \begin{pmatrix} 1 \\ -2 \\ 0 \end{pmatrix} e^{-t}.$$

44. For $\mathbf{X} = t^{\lambda}\mathbf{K}$ we have $\mathbf{X}' = \lambda t^{\lambda-1}\mathbf{K}$. Substituting into the differential equation, we have

$$\lambda t^{\lambda}\mathbf{K} = \begin{pmatrix} 2 & -2 \\ 2 & 7 \end{pmatrix} t^{\lambda}\mathbf{K}.$$

Since $t > 0$, we have

$$\lambda\mathbf{K} = \begin{pmatrix} 2 & -2 \\ 2 & 7 \end{pmatrix} \mathbf{K}$$

or

$$\left[\begin{pmatrix} 2 & -2 \\ 2 & 7 \end{pmatrix} - \lambda I\right]\mathbf{K} = 0.$$

Thus, λ is an eigenvalue for the matrix and $\mathbf{K}$ is the corresponding eigenvector. The characteristic equation is

$$\begin{vmatrix} 2-\lambda & -2 \\ 2 & 7-\lambda \end{vmatrix} = 0$$

or $(\lambda-3)(\lambda-6)=0$. The eigenvalues are $\lambda_1 = 3$ and $\lambda_2 = 6$. For $\lambda_1 = 3$ we have

$$\begin{aligned} -k_1 - 2k_2 &= 0 \\ 2k_1 + 4k_2 &= 0. \end{aligned}$$

An eigenvector corresponding to $\lambda_1 = 3$ is

$$\mathbf{K}_1 = \begin{pmatrix} 2 \\ -1 \end{pmatrix}.$$

For $\lambda_2 = 6$ we have

$$\begin{aligned} -4k_1 - 2k_2 &= 0 \\ 2k_1 + k_2 &= 0. \end{aligned}$$

An eigenvector corresponding to $\lambda_2 = 6$ is

$$\mathbf{K}_2 = \begin{pmatrix} 1 \\ -2 \end{pmatrix}.$$

A family of solutions is

$$\mathbf{X}(t) = c_1 t^3 \begin{pmatrix} 2 \\ -1 \end{pmatrix} + c_2 t^6 \begin{pmatrix} 1 \\ -2 \end{pmatrix}.$$

Exercises 8.7

4. To solve the system

$$\begin{aligned} \frac{dx}{dt} &= x - 4y + 4t + 9e^{6t} \\ \frac{dy}{dt} &= 4x + y - t + e^{6t} \end{aligned}$$

we first solve the related homogeneous system

$$\begin{aligned} \frac{dx}{dt} &= x - 4y \\ \frac{dy}{dt} &= 4x + y. \end{aligned}$$

The eigenvalues are determined from

$$\begin{vmatrix} 1-\lambda & -4 \\ 4 & 1-\lambda \end{vmatrix} = \lambda^2 - 2\lambda + 17 = 0.$$

Thus, $\lambda_1 = 1+4i$ and $\lambda_2 = 1-4i$. An eigenvector corresponding to $\lambda_1 = 1+4i$ is found to be

$$\mathbf{K}_1 = \begin{pmatrix} i \\ 1 \end{pmatrix}.$$

Thus, an eigenvector corresponding to $\lambda_2 = 1-4i$ is

$$\mathbf{K}_2 = \begin{pmatrix} -i \\ 1 \end{pmatrix}.$$

Consequently,

$$\begin{aligned}\mathbf{X}_c &= c_1\left[\begin{pmatrix} 0 \\ 1 \end{pmatrix}\cos 4t + \begin{pmatrix} -1 \\ 0 \end{pmatrix}\sin 4t\right]e^t + c_2\left[\begin{pmatrix} -1 \\ 0 \end{pmatrix}\cos 4t - \begin{pmatrix} 0 \\ 1 \end{pmatrix}\sin 4t\right]e^t \\ &= c_1\begin{pmatrix} -\sin 4t \\ \cos 4t \end{pmatrix}e^t + c_2\begin{pmatrix} -\cos 4t \\ -\sin 4t \end{pmatrix}e^t.\end{aligned}$$

Now by writing the original system as

$$\mathbf{X}' = \begin{pmatrix} 1 & -4 \\ 4 & 1 \end{pmatrix}\mathbf{X} + \begin{pmatrix} 4 \\ -1 \end{pmatrix}t + \begin{pmatrix} 9 \\ 1 \end{pmatrix}e^{6t}$$

we are prompted to seek a particular solution of the form

$$\mathbf{X}_p = \begin{pmatrix} a_3 \\ b_3 \end{pmatrix} + \begin{pmatrix} a_2 \\ b_2 \end{pmatrix} + \begin{pmatrix} a_1 \\ b_1 \end{pmatrix}e^{6t}.$$

Substituting $\mathbf{X}_p$ into the system gives

$$\begin{pmatrix} a_3 \\ b_3 \end{pmatrix} + \begin{pmatrix} 6a_1 \\ 6b_1 \end{pmatrix}e^{6t} = \begin{pmatrix} 1 & -4 \\ 4 & 1 \end{pmatrix}\left[\begin{pmatrix} a_3 \\ b_3 \end{pmatrix}t + \begin{pmatrix} a_2 \\ b_2 \end{pmatrix} + \begin{pmatrix} a_1 \\ b_1 \end{pmatrix}e^{6t}\right] + \begin{pmatrix} 4 \\ -1 \end{pmatrix}t + \begin{pmatrix} 9 \\ 1 \end{pmatrix}e^{6t}$$

or

$$\begin{pmatrix} a_3 \\ b_3 \end{pmatrix} + \begin{pmatrix} 6a_1 \\ 6b_1 \end{pmatrix}e^{6t} = \begin{pmatrix} a_3 - 4b_3 \\ 4a_3 + b_3 \end{pmatrix}t + \begin{pmatrix} a_2 - 4b_2 \\ 4a_2 + b_2 \end{pmatrix} + \begin{pmatrix} a_1 - 4b_1 \\ 4a_1 + b_1 \end{pmatrix}e^{6t} + \begin{pmatrix} 4 \\ -1 \end{pmatrix}t + \begin{pmatrix} 9 \\ 1 \end{pmatrix}e^{6t}.$$

Then

$$\begin{pmatrix} 0 \\ 0 \end{pmatrix} = \begin{pmatrix} (a_3 - 4b_3 + 4)t + (a_2 - 4b_2 - a_3) + (-5a_1 - 4b_1 + 9)e^{6t} \\ (4a_3 + b_3 - 1)t + (4a_2 + b_2 - b_3) + (4a_1 - 5b_1 + 1)e^{6t} \end{pmatrix}$$

and so

$$a_3 - 4b_3 + 4 = 0$$

$$4a_3 + b_3 - 1 = 0$$

$$a_2 - 4b_2 - a_3 = 0$$

$$4a_2 + b_2 - b_3 = 0$$

$$-5a_1 - 4b_1 + 9 = 0$$

$$4a_1 - 5b_1 + 1 = 0.$$

From the first two equations, we find $a_3 = 0$ and $b_3 = 1$. Substituting these values into the next two equations and solving then yields $a_2 = 4/17$ and $b_2 = 1/17$. The last two equations give $a_1 = 1$, $b_1 = 1$. It follows that

$$\mathbf{X}_p = \begin{pmatrix} 0 \\ 1 \end{pmatrix} t + \begin{pmatrix} \frac{4}{17} \\ \frac{1}{17} \end{pmatrix} + \begin{pmatrix} 1 \\ 1 \end{pmatrix} e^{6t}.$$

The general solution of the system is then

$$\begin{aligned} \mathbf{X} &= \mathbf{X}_c + \mathbf{X}_p \\ &= c_1 \begin{pmatrix} -\sin 4t \\ \cos 4t \end{pmatrix} e^t + c_2 \begin{pmatrix} -\cos 4t \\ -\sin 4t \end{pmatrix} e^t + \begin{pmatrix} 0 \\ 1 \end{pmatrix} t + \frac{1}{17} \begin{pmatrix} 4 \\ 1 \end{pmatrix} + \begin{pmatrix} 1 \\ 1 \end{pmatrix} e^{6t}. \end{aligned}$$

6. The eigenvalues of $\begin{pmatrix} -1 & 5 \\ -1 & 1 \end{pmatrix}$ are $\lambda_1 = 2i$ and $\lambda_2 = -2i$. The corresponding eigenvectors are found to be

$$\mathbf{K}_1 = \begin{pmatrix} 5 \\ 1 + 2i \end{pmatrix} \quad \text{and} \quad \mathbf{K}_2 = \begin{pmatrix} 5 \\ 1 - 2i \end{pmatrix}.$$

From these we obtain

$$\mathbf{X}_c = c_1 \begin{pmatrix} 5\cos 2t \\ \cos 2t - 2\sin 2t \end{pmatrix} + c_2 \begin{pmatrix} 5\sin 2t \\ 2\cos 2t + \sin 2t \end{pmatrix}.$$

Now, by writing the system as

$$\mathbf{X}' = \begin{pmatrix} -1 & 5 \\ -1 & 1 \end{pmatrix} \mathbf{X} + \begin{pmatrix} 0 \\ -2 \end{pmatrix} \cos t + \begin{pmatrix} 1 \\ 0 \end{pmatrix} \sin t,$$

we are inspired to try

$$\mathbf{X}_p = \begin{pmatrix} a_2 \\ b_2 \end{pmatrix} \cos t + \begin{pmatrix} a_1 \\ b_1 \end{pmatrix} \sin t$$

as a particular solution. Substituting $\mathbf{X}_p$ into the system then gives, after simplifying,

$$\begin{pmatrix} 0 \\ 0 \end{pmatrix} = \begin{pmatrix} (-a_2 + 5b_2 - a_1)\cos t + (-a_1 + 5b_1 + a_2 + 1)\sin t \\ (-a_2 + b_2 - b_1 - 2)\cos t + (-a_1 + b_1 + b_2)\sin t \end{pmatrix}.$$

Consequently,

$$-a_2 + 5b_2 - a_1 = 0$$
$$-a_2 + b_2 - b_1 - 2 = 0$$
$$-a_1 + 5b_1 + a_2 + 1 = 0$$
$$-a_1 + b_1 + b_2 = 0.$$

A solution of this system of equations is found to be $a_2 = -3$, $b_2 = -2/3$, $a_1 = -1/3$, $b_1 = 1/3$. Thus

$$\mathbf{X}_p = \begin{pmatrix} -3 \\ -\frac{2}{3} \end{pmatrix} \cos t + \begin{pmatrix} -\frac{1}{3} \\ \frac{1}{3} \end{pmatrix} \sin t$$

and

$$\mathbf{X} = c_1 \begin{pmatrix} 5\cos 2t \\ \cos 2t - 2\sin 2t \end{pmatrix} + c_2 \begin{pmatrix} 5\sin 2t \\ 2\cos 2t + \sin 2t \end{pmatrix} + \begin{pmatrix} -3 \\ -\frac{2}{3} \end{pmatrix} \cos t + \begin{pmatrix} -\frac{1}{3} \\ \frac{1}{3} \end{pmatrix} \sin t.$$

Exercises 8.8

6. To solve the system

$$\mathbf{X}' = \begin{pmatrix} 0 & 2 \\ -1 & 3 \end{pmatrix} \mathbf{X} + \begin{pmatrix} 2 \\ e^{-3t} \end{pmatrix}$$

we first solve the related homogeneous system

$$\mathbf{X}' = \begin{pmatrix} 0 & 2 \\ -1 & 3 \end{pmatrix} \mathbf{X}.$$

The eigenvalues are determined from

$$\begin{vmatrix} -\lambda & 2 \\ -1 & 3-\lambda \end{vmatrix} = \lambda^2 - 3\lambda + 2 = (\lambda - 1)(\lambda - 2) = 0.$$

Thus, the eigenvalues are $\lambda_1 = 1$ and $\lambda_2 = 2$. An eigenvector corresponding to $\lambda_1 = 1$ is found from

$$-k_1 + 2k_2 = 0$$
$$-k_1 + 2k_2 = 0.$$

Choosing $k_2 = 1$ we se that

$$\mathbf{K}_1 = \begin{pmatrix} 2 \\ 1 \end{pmatrix}$$

is an eigenvector corresponding to $\lambda_1 = 1$. For $\lambda_2 = 2$ we have

$$\begin{aligned} -2k_1 + 2k_2 &= 0 \\ -k_1 + k_2 &= 0, \end{aligned}$$

so that

$$\mathbf{K}_2 = \begin{pmatrix} 1 \\ 1 \end{pmatrix}$$

is an eigenvector corresponding to $\lambda_2 = 2$. Thus, the fundamental matrix of the related homogeneous system is

$$\mathbf{\Phi}(t) = \begin{pmatrix} 2e^t & e^{2t} \\ e^t & e^{2t} \end{pmatrix}.$$

Now,

$$\det \mathbf{\Phi}(t) = \begin{vmatrix} 2e^t & e^{2t} \\ e^t & e^{2t} \end{vmatrix} = 2e^{3t} - e^{3t} = e^{3t}$$

so

$$\mathbf{\Phi}^{-1}(t) = \frac{1}{e^{3t}} \begin{pmatrix} e^{2t} & -e^{2t} \\ -e^t & 2e^t \end{pmatrix} = \begin{pmatrix} e^{-t} & -e^{-t} \\ -e^{-2t} & 2e^{-2t} \end{pmatrix}.$$

We identify

$$\mathbf{F}(t) = \begin{pmatrix} 2 \\ e^{-3t} \end{pmatrix},$$

so

$$\mathbf{\Phi}^{-1}(t)\mathbf{F}(t) = \begin{pmatrix} e^{-t} & -e^{-t} \\ -e^{-2t} & 2e^{-2t} \end{pmatrix} \begin{pmatrix} 2 \\ e^{-3t} \end{pmatrix} = \begin{pmatrix} 2e^{-t} - e^{-4t} \\ -2e^{-2t} + 2e^{-5t} \end{pmatrix}$$

and

$$\int \mathbf{\Phi}^{-1}(t)\mathbf{F}(t)\,dt = \int \begin{pmatrix} 2e^{-t} - e^{-4t} \\ -2e^{-2t} + 2e^{-5t} \end{pmatrix} dt = \begin{pmatrix} -2e^{-t} + \frac{1}{4}e^{-4t} \\ e^{-2t} - \frac{2}{5}e^{-5t} \end{pmatrix}.$$

Then

$$\begin{aligned} \mathbf{X}_p &= \mathbf{\Phi}(t) \int \mathbf{\Phi}^{-1}(t)\mathbf{F}(t)\,dt \\ &= \begin{pmatrix} 2e^t & e^{2t} \\ e^t & e^{2t} \end{pmatrix} \begin{pmatrix} -2e^{-t} + \frac{1}{4}e^{-4t} \\ e^{-2t} - \frac{2}{5}e^{-5t} \end{pmatrix} \\ &= \begin{pmatrix} -4 + \frac{1}{2}e^{-3t} + 1 - \frac{2}{5}e^{-3t} \\ -2 + \frac{1}{4}e^{-3t} + 1 - \frac{2}{5}e^{-3t} \end{pmatrix} \\ &= \begin{pmatrix} \frac{1}{10}e^{-3t} - 3 \\ -\frac{3}{20}e^{-3t} - 1 \end{pmatrix} = \frac{1}{20} \begin{pmatrix} 2 \\ -3 \end{pmatrix} e^{-3t} - \begin{pmatrix} 3 \\ 1 \end{pmatrix}. \end{aligned}$$

Hence the general solution of the original system is

$$\mathbf{X}(t) = c_1 \begin{pmatrix} 2 \\ 1 \end{pmatrix} e^t + c_2 \begin{pmatrix} 1 \\ 1 \end{pmatrix} e^{2t} + \frac{1}{20} \begin{pmatrix} 2 \\ -3 \end{pmatrix} e^{-3t} - \begin{pmatrix} 3 \\ 1 \end{pmatrix}.$$

10. To solve the system

$$\mathbf{X}' = \begin{pmatrix} 3 & 2 \\ -2 & -1 \end{pmatrix} \mathbf{X} + \begin{pmatrix} 1 \\ 1 \end{pmatrix}$$

we first solve the related homogeneous system

$$\mathbf{X}' = \begin{pmatrix} 3 & 2 \\ -2 & -1 \end{pmatrix} \mathbf{X}.$$

The eigenvalues are determined from

$$\begin{vmatrix} 3-\lambda & 2 \\ -2 & -1-\lambda \end{vmatrix} = \lambda^2 - 2\lambda + 1 = (\lambda - 1)^2 = 0.$$

Thus, $\lambda_1 = 1$ is an eigenvalue of multiplicity two. An eigenvector corresponding to $\lambda_1 = 1$ is found from

$$2k_1 + 2k_2 = 0$$

$$-2k_1 - 2k_2 = 0.$$

Choosing $k_1 = 1$ we see that $k_2 = -1$ and so

$$\mathbf{K} = \begin{pmatrix} 1 \\ -1 \end{pmatrix}$$

is an eigenvector. Hence, one solution of the related homogeneous system is

$$\mathbf{X}_1 = \begin{pmatrix} 1 \\ -1 \end{pmatrix} e^t.$$

Now a second solution can be found of the form

$$\mathbf{X}_2 = \mathbf{K}te^t + \mathbf{P}e^t.$$

To find the vector $\mathbf{P}$ we need only solve

$$(\mathbf{A} - \mathbf{I})\mathbf{P} = \mathbf{K} \quad \text{or} \quad \begin{pmatrix} 2 & 2 \\ -2 & -2 \end{pmatrix} \begin{pmatrix} p_1 \\ p_2 \end{pmatrix} = \begin{pmatrix} 1 \\ -1 \end{pmatrix}.$$

Multiplying out this last expression gives

$$2p_1 + 2p_2 = 1$$

$$-2p_1 - 2p_2 = -1.$$

Choosing $p_1 = 0$ gives $p_2 = 1/2$. Thus $\mathbf{P} = \begin{pmatrix} 0 \\ \frac{1}{2} \end{pmatrix}$ and so

$$\mathbf{X}_2 = \begin{pmatrix} 1 \\ -1 \end{pmatrix} te^t + \begin{pmatrix} 0 \\ \frac{1}{2} \end{pmatrix} e^t = \begin{pmatrix} t \\ \frac{1}{2} - t \end{pmatrix} e^t.$$

The fundamental matrix of the homogeneous system is

$$\mathbf{\Phi}(t) = \begin{pmatrix} e^t & te^t \\ -e^t & \frac{1}{2}e^t - te^t \end{pmatrix}.$$

It follows that

$$\det \mathbf{\Phi}(t) = \begin{vmatrix} e^t & te^t \\ -e^t & \frac{1}{2}e^t - te^t \end{vmatrix} = \frac{1}{2}e^{2t} - te^{2t} + te^{2t} = \frac{1}{2}e^{2t},$$

and

$$\mathbf{\Phi}^{-1}(t) = \frac{2}{e^{2t}} \begin{pmatrix} \frac{1}{2}e^t - te^t & -te^t \\ e^t & e^t \end{pmatrix} = \begin{pmatrix} e^{-t} - 2te^{-t} & -2te^{-t} \\ 2e^{-t} & 2e^{-t} \end{pmatrix}.$$

Now we identify

$$\mathbf{F}(t) = \begin{pmatrix} 1 \\ 1 \end{pmatrix},$$

so

$$\mathbf{\Phi}^{-1}(t)\mathbf{F}(t) = \begin{pmatrix} e^{-t} - 2te^{-t} & -2te^{-t} \\ 2e^{-t} & 2e^{-t} \end{pmatrix} \begin{pmatrix} 1 \\ 1 \end{pmatrix} = \begin{pmatrix} e^{-t} - 4te^{-t} \\ 4e^{-t} \end{pmatrix},$$

and

$$\int \mathbf{\Phi}^{-1}(t)\mathbf{F}(t)\,dt = \int \begin{pmatrix} e^{-t} - 4te^{-t} \\ 4e^{-t} \end{pmatrix} dt = \begin{pmatrix} -e^{-t} + 4te^{-t} + 4e^{-t} \\ -4e^{-t} \end{pmatrix} = \begin{pmatrix} 3e^{-t} + 4te^{-t} \\ -4e^{-t} \end{pmatrix}.$$

Then,

$$\begin{aligned} \mathbf{X}_p &= \mathbf{\Phi}(t) \int \mathbf{\Phi}^{-1}(t)\mathbf{F}(t)\,dt \\ &= \begin{pmatrix} e^t & te^t \\ -e^t & \frac{1}{2}e^t - te^t \end{pmatrix} \begin{pmatrix} 3e^{-t} + 4te^{-t} \\ -4e^{-t} \end{pmatrix} \\ &= \begin{pmatrix} 3 + 4t - 4t \\ -3 - 4t - 2 + 4t \end{pmatrix} = \begin{pmatrix} 3 \\ -5 \end{pmatrix} \end{aligned}$$

and the general solution of the original system is

$$\mathbf{X}(t) = c_1 \begin{pmatrix} 1 \\ -1 \end{pmatrix} e^t + c_2 \begin{pmatrix} t \\ \frac{1}{2} - t \end{pmatrix} e^t + \begin{pmatrix} 3 \\ -5 \end{pmatrix}.$$

22. To solve the system

$$\mathbf{X}' = \begin{pmatrix} 1 & -1 \\ 1 & -1 \end{pmatrix} \mathbf{X} + \begin{pmatrix} \frac{1}{t} \\ \frac{1}{t} \end{pmatrix}$$

we first solve the related homogeneous system

$$\mathbf{X}' = \begin{pmatrix} 1 & -1 \\ 1 & -1 \end{pmatrix} \mathbf{X}.$$

The eigenvalues are determined from

$$\begin{vmatrix} 1-\lambda & -1 \\ 1 & -1-\lambda \end{vmatrix} = \lambda^2 = 0.$$

Thus, the eigenvalues are $\lambda_1 = \lambda_2 = 0$. An eigenvector corresponding to $\lambda_1 = 0$ is found from

$$k_1 - k_2 = 0$$

$$k_1 - k_2 = 0.$$

Choosing $k_1 = 1$ we see that $k_2 = 1$ and

$$\mathbf{K} = \begin{pmatrix} 1 \\ 1 \end{pmatrix}$$

is an eigenvector. Hence, one solution of the related homogeneous system is

$$\mathbf{X}_1 = \begin{pmatrix} 1 \\ 1 \end{pmatrix}.$$

Since $\lambda_1 = 0$ is an eigenvalue of multiplicity two a second solution can be found of the form

$$\mathbf{X}_2 = \mathbf{K}te^{0t} + \mathbf{P}e^{0t} = \mathbf{K}t + \mathbf{P}.$$

The vector $\mathbf{P}$ is found from

$$(\mathbf{A} - 0\mathbf{I})\mathbf{P} = \mathbf{K} \quad \text{or} \quad \begin{pmatrix} 1 & -1 \\ 1 & -1 \end{pmatrix} \begin{pmatrix} p_1 \\ p_2 \end{pmatrix} = \begin{pmatrix} 1 \\ 1 \end{pmatrix}.$$

That is

$$p_1 - p_2 = 1$$

$$p_1 - p_2 = 1.$$

By choosing $p_1 = 1$ we get $p_2 = 0$. Consequently,

$$\mathbf{X}_2 = \begin{pmatrix} 1 \\ 1 \end{pmatrix} t + \begin{pmatrix} 1 \\ 0 \end{pmatrix} = \begin{pmatrix} t+1 \\ t \end{pmatrix},$$

and the fundamental matrix of the homogeneous system is

$$\mathbf{\Phi}(t) = \begin{pmatrix} 1 & t+1 \\ 1 & t \end{pmatrix}.$$

Now,

$$\det \mathbf{\Phi}(t) = \begin{vmatrix} 1 & t+1 \\ 1 & t \end{vmatrix} = -1,$$

so

$$\mathbf{\Phi}^{-1}(t) = -\begin{pmatrix} t & -t-1 \\ -1 & 1 \end{pmatrix} = \begin{pmatrix} -t & t+1 \\ 1 & -1 \end{pmatrix}.$$

Now we identify $t_0 = 1$,

$$\mathbf{X}_0 = \begin{pmatrix} 2 \\ -1 \end{pmatrix},$$

and

$$\mathbf{F}(t) = \begin{pmatrix} \frac{1}{t} \\ \frac{1}{t} \end{pmatrix}.$$

From equation (11) in the text then,

$$\begin{aligned}
\mathbf{X}(t) &= \mathbf{\Phi}(t)\mathbf{\Phi}^{-1}(t_0)\mathbf{X}_0 + \mathbf{\Phi}(t)\int_{t_0}^{t} \mathbf{\Phi}^{-1}(s)\mathbf{F}(s)\,ds \\
&= \begin{pmatrix} 1 & t+1 \\ 1 & t \end{pmatrix}\begin{pmatrix} -1 & 2 \\ 1 & -1 \end{pmatrix}\begin{pmatrix} 2 \\ -1 \end{pmatrix} + \begin{pmatrix} 1 & t+1 \\ 1 & t \end{pmatrix}\int_1^t \begin{pmatrix} -s & s+1 \\ 1 & -1 \end{pmatrix}\begin{pmatrix} \frac{1}{s} \\ \frac{1}{s} \end{pmatrix} ds \\
&= \begin{pmatrix} t & 1-t \\ -1+t & 2-t \end{pmatrix}\begin{pmatrix} 2 \\ -1 \end{pmatrix} + \begin{pmatrix} 1 & t+1 \\ 1 & t \end{pmatrix}\int_1^t \begin{pmatrix} \frac{1}{s} \\ 0 \end{pmatrix} ds \\
&= \begin{pmatrix} 3t-1 \\ 3t-4 \end{pmatrix} + \begin{pmatrix} 1 & t+1 \\ 1 & t \end{pmatrix}\begin{pmatrix} \ln s \\ 0 \end{pmatrix}\Bigg|_1^t \\
&= \begin{pmatrix} 3t-1 \\ 3t-4 \end{pmatrix} + \begin{pmatrix} 1 & t+1 \\ 1 & t \end{pmatrix}\begin{pmatrix} \ln t \\ 0 \end{pmatrix} \\
&= \begin{pmatrix} 3t-1 \\ 3t-4 \end{pmatrix} + \begin{pmatrix} \ln t \\ \ln t \end{pmatrix} \\
&= \begin{pmatrix} 3 \\ 3 \end{pmatrix} t - \begin{pmatrix} 1 \\ 4 \end{pmatrix} + \begin{pmatrix} 1 \\ 1 \end{pmatrix} \ln t.
\end{aligned}$$

Exercises 8.9

2. For

$$\mathbf{A} = \begin{pmatrix} 1 & 0 \\ 0 & 2 \end{pmatrix}$$

we have

$$\mathbf{A}^2 = \begin{pmatrix} 1 & 0 \\ 0 & 2 \end{pmatrix}\begin{pmatrix} 1 & 0 \\ 0 & 2 \end{pmatrix} = \begin{pmatrix} 1 & 0 \\ 0 & 4 \end{pmatrix},$$

$$\mathbf{A}^3 = \mathbf{A}\mathbf{A}^2 = \begin{pmatrix} 1 & 0 \\ 0 & 2 \end{pmatrix}\begin{pmatrix} 1 & 0 \\ 0 & 4 \end{pmatrix} = \begin{pmatrix} 1 & 0 \\ 0 & 8 \end{pmatrix},$$

$$\mathbf{A}^4 = \mathbf{A}\mathbf{A}^3 = \begin{pmatrix} 1 & 0 \\ 0 & 2 \end{pmatrix}\begin{pmatrix} 1 & 0 \\ 0 & 8 \end{pmatrix} = \begin{pmatrix} 1 & 0 \\ 0 & 16 \end{pmatrix},$$

and so on. In general

$$\mathbf{A}^k = \begin{pmatrix} 1 & 0 \\ 0 & 2^k \end{pmatrix} \quad \text{for} \quad k = 1, 2, 3, \ldots.$$

Thus,

$$\begin{aligned} e^{t\mathbf{A}} &= \mathbf{I} + \frac{\mathbf{A}}{1!}t + \frac{\mathbf{A}^2}{2!}t^2 + \frac{\mathbf{A}^3}{3!}t^3 + \cdots \\ &= \begin{pmatrix} 1 & 0 \\ 0 & 1 \end{pmatrix} + \frac{1}{1!}\begin{pmatrix} 1 & 0 \\ 0 & 2 \end{pmatrix}t + \frac{1}{2!}\begin{pmatrix} 1 & 0 \\ 0 & 4 \end{pmatrix}t^2 + \frac{1}{3!}\begin{pmatrix} 1 & 0 \\ 0 & 8 \end{pmatrix}t^3 + \cdots \\ &= \begin{pmatrix} 1 + t + \frac{t^2}{2!} + \frac{t^3}{3!} + \cdots & 0 \\ 0 & 1 + t + \frac{(2t)^2}{2!} + \frac{(2t)^3}{3!} + \cdots \end{pmatrix} \\ &= \begin{pmatrix} e^t & 0 \\ 0 & e^{2t} \end{pmatrix} \end{aligned}$$

and

$$e^{-t\mathbf{A}} = \begin{pmatrix} e^{-t} & 0 \\ 0 & e^{-2t} \end{pmatrix}.$$

8. To solve

$$\mathbf{X}' = \begin{pmatrix} 1 & 0 \\ 0 & 2 \end{pmatrix}\mathbf{X} + \begin{pmatrix} 3 \\ -1 \end{pmatrix}$$

we identify $t_0 = 0$,

$$\mathbf{F}(s) = \begin{pmatrix} 3 \\ -1 \end{pmatrix},$$

and use the results of Problem 2 and equation (3) in the text.

$$\begin{aligned}
\mathbf{X}(t) &= e^{t\mathbf{A}}\mathbf{C} + e^{t\mathbf{A}}\int_{t_0}^{t} e^{-s\mathbf{A}}\mathbf{F}(s)\,ds \\
&= \begin{pmatrix} e^t & 0 \\ 0 & e^{2t} \end{pmatrix}\begin{pmatrix} c_1 \\ c_2 \end{pmatrix} + \begin{pmatrix} e^t & 0 \\ 0 & e^{2t} \end{pmatrix}\int_0^t \begin{pmatrix} e^{-s} & 0 \\ 0 & e^{-2s} \end{pmatrix}\begin{pmatrix} 3 \\ -1 \end{pmatrix} ds \\
&= \begin{pmatrix} c_1e^t \\ c_2e^{2t} \end{pmatrix} + \begin{pmatrix} e^t & 0 \\ 0 & e^{2t} \end{pmatrix}\int_0^t \begin{pmatrix} 3e^{-s} \\ -e^{-2s} \end{pmatrix} ds \\
&= \begin{pmatrix} c_1e^t \\ c_2e^{2t} \end{pmatrix} + \begin{pmatrix} e^t & 0 \\ 0 & e^{2t} \end{pmatrix}\begin{pmatrix} -3e^{-s} \\ \frac{1}{2}e^{-2s} \end{pmatrix}\Bigg|_0^t \\
&= \begin{pmatrix} c_1e^t \\ c_2e^{2t} \end{pmatrix} + \begin{pmatrix} e^t & 0 \\ 0 & e^{2t} \end{pmatrix}\begin{pmatrix} -3e^{-t} - 3 \\ \frac{1}{2}e^{-2t} - \frac{1}{2} \end{pmatrix} \\
&= \begin{pmatrix} c_1e^t \\ c_2e^{2t} \end{pmatrix} + \begin{pmatrix} -3 - 3e^t \\ \frac{1}{2} - \frac{1}{2}e^{2t} \end{pmatrix} \\
&= c_1\begin{pmatrix} 1 \\ 0 \end{pmatrix}e^t + c_2\begin{pmatrix} 0 \\ 1 \end{pmatrix}e^{2t} + \begin{pmatrix} -3 \\ \frac{1}{2} \end{pmatrix} - \begin{pmatrix} 3e^t \\ \frac{1}{2}e^{2t} \end{pmatrix} \\
&= \begin{pmatrix} c_1e^t - 3e^t \\ c_2e^{2t} - \frac{1}{2}e^{2t} \end{pmatrix} + \begin{pmatrix} -3 \\ \frac{1}{2} \end{pmatrix} \\
&= c_3\begin{pmatrix} 1 \\ 0 \end{pmatrix}e^t + c_4\begin{pmatrix} 0 \\ 1 \end{pmatrix}e^{2t} + \begin{pmatrix} -3 \\ \frac{1}{2} \end{pmatrix}.
\end{aligned}$$

9 Numerical Methods

Exercises 9.1

6. The isoclines for the differential equation

$$y' = \left(x^2 + y^2\right)^{-1}$$

are the curves

$$\left(x^2 + y^2\right)^{-1} = c \qquad \text{or} \qquad x^2 + y^2 = \frac{1}{c}$$

for $c > 0$. Thus, the isoclines are a family of concentric circles centered at the origin.

12. The isoclines for the differential equation $y' = x + y$ are $x + y = c$ or $y = -x+c$. This is a family of straight lines with slopes -1 and y-intercepts c. We note that when $c = -1$ the lineal elements are co-linear with the line $y = -x - 1$. This line is a particular solution of the differential equation. The direction field is shown in the figure along with some possible solution curves.

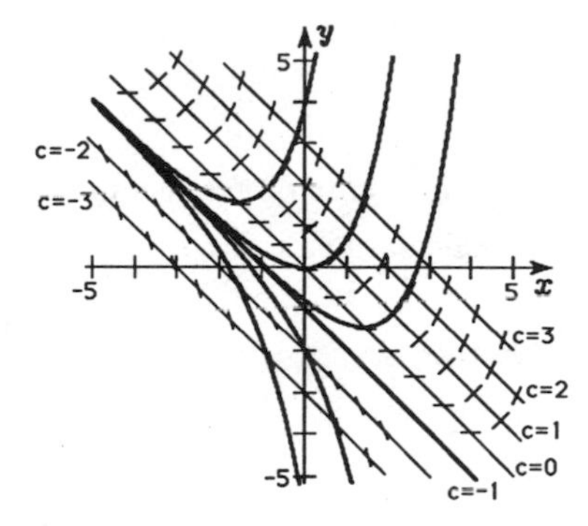

24. The isoclines of the differential equation

$$\frac{dy}{dx} = \frac{2y}{x+y}$$

are the straight lines

$$\frac{2y}{x+y} = c \qquad \text{or} \qquad y = \frac{c}{2-c}x.$$

A line in this family will be a solution of the differential equation whenever its slope is c; that is, we want $c/(2-c) = c$ or $c = 1$. Therefore, the isocline $y = x$ is also a solution of the differential equation.

Exercises 9.2

6. In the initial value problem

$$y' = x^2 + y^2, \quad y(0) = 1$$

we identify $f(x,y) = x^2 + y^2$. The iteration formula is

$$y_{n+1} = y_n + h\left(x_n^2 + y_n^2\right).$$

For $h = 0.1$ we find

$$\begin{aligned} y_1 &= y_0 + 0.1\left(x_0^2 + y_0^2\right) \\ &= 1 + 0.1(0+1) \\ &= 1.1. \end{aligned}$$

To find y_2 we compute

$$\begin{aligned} y_2 &= y_1 + 0.1\left(x_1^2 + y_1^2\right) \\ &= 1.1 + 0.1(0.01 + 1.21) \\ &= 1.222. \end{aligned}$$

The results of the remaining calculations are shown in the table. Also shown are the results of the calculations using $h = 0.5$.

Euler's Method with $h = 0.1$	
x_n	y_n
0.00	1.0000
0.10	1.1000
0.20	1.2220
0.30	1.3753
0.40	1.5735
0.50	1.8371

Euler's Method with $h = 0.05$	
x_n	y_n
0.00	1.0000
0.05	1.0500
0.10	1.1053
0.15	1.1668
0.20	1.2360
0.25	1.3144
0.30	1.4039
0.35	1.5070
0.40	1.6267
0.45	1.7670
0.50	1.9332

12. In the initial value problem

$$y' = y - y^2, \quad y(0) = 0.5$$

we identify $f(x, y) = y - y^2$. The iteration formula is

$$y_{n+1} = y_n + h\left(y_n - y_n^2\right).$$

For $h = 0.1$ we find

$$\begin{aligned} y_1 &= y_0 + 0.1\left(y_0 - y_0^2\right) \\ &= 0.5 + 0.1(0.5 - 0.25) \\ &= 0.525. \end{aligned}$$

To find y_2 we compute

$$\begin{aligned} y_2 &= y_1 + 0.1\left(y_1 - y_1^2\right) \\ &= 0.525 + 0.1(0.525 - 0.2756) \\ &= 0.5499. \end{aligned}$$

The results of the remaining calculations are shown in the table. Also shown are the results of the calculations using $h = 0.05$.

Euler's Method with $h = 0.1$	
x_n	y_n
0.00	0.5000
0.10	0.5250
0.20	0.5499
0.30	0.5747
0.40	0.5991
0.50	0.6231

Euler's Method with $h = 0.05$	
x_n	y_n
0.00	0.5000
0.05	0.5125
0.10	0.5250
0.15	0.5375
0.20	0.5499
0.25	0.5623
0.30	0.5746
0.35	0.5868
0.40	0.5989
0.45	0.6109
0.50	0.6228

16. For the initial value problem

$$y' = 2xy, \quad y(1) = 1$$

we identify $f(x, y) = 2xy$. Then the iteration formula for $y_{1,k+1}$ is

$$y_{1,k+1} = y_0 + h\left(\frac{2x_0y_0 + 2x_1y_{1,k}}{2}\right).$$

Now $x_0 = 1$, $y_0 = 1$, and $h = 0.1$ So

$$\begin{aligned} y_{1,k+1} &= 1 + 0.1\left(1 + 1.1y_{1,k}\right) \\ &= 1.1 + 0.11y_{1,k}. \end{aligned}$$

Taking

$$\begin{aligned} y_{1,0} &= y_1 \\ &= y_0 + hf(x_0, y_0) \\ &= 1 + 0.1(2 \cdot 1 \cdot 1) \\ &= 1.2 \end{aligned}$$

we compute

$$\begin{aligned} y_{1,1} &= 1.1 + 0.11(1.2) \\ &= 1.2320. \end{aligned}$$

From Table 9.1 in the text we see that the true value of y_1 is 1.2337. Thus, the percentage relative error is

$$E_{1,1} = \frac{|1.2337 - 1.2320|}{1.2337} \times 100 = 0.14.$$

The remaining approximations and percentage relative errors are given in the table.

k	$y_{1,k}$	$E_{1,k}$
0	1.2000	2.73
1	1.2320	0.14
2	1.2355	0.15
3	1.2359	0.18
4	1.2359	0.18
5	1.2360	0.19

Exercises 9.3

4. The initial value problem is

$$y' = x^2 + y^2, \quad y(0) = 1,$$

so

$$\begin{aligned} y'' &= 2x + 2yy' \\ &= 2x + 2y\left(x^2 + y^2\right) \\ &= 2x + 2x^2y + 2y^3. \end{aligned}$$

The iteration formula is then

$$\begin{aligned} y_{n+1} &= y_n + \left(x_n^2 + y_n^2\right)h + \left(2x_n + 2x_n^2 + 2y_n^3\right)\frac{h^2}{2} \\ &= y_n + \left(x_n^2 + y_n^2\right)h + \left(x_n + x_n^2 y_n + y_n^3\right)h^2. \end{aligned}$$

For $h = 0.1$ we find

$$\begin{aligned} y_1 &= y_0 + \left(x_0^2 + y_0^2\right)(0.1) + \left(x_0 + x_0^2 y_0 + y_0^3\right)(0.01) \\ &= 1 + (0 + 1)(0.1) + (0 + 0 + 1)(0.01) \\ &= 1.11. \end{aligned}$$

The results of the remaining calculations are shown in the table. Also shown is the table obtained when $h = 0.5$.

3-Term Taylor with $h = 0.1$

x_n	y_n
0.00	1.0000
0.10	1.1100
0.20	1.2490
0.30	1.4310
0.40	1.6783
0.50	2.0300

3-Term Taylor with $h = 0.05$

x_n	y_n
0.00	1.0000
0.05	1.0525
0.10	1.1111
0.15	1.1770
0.20	1.2519
0.25	1.3378
0.30	1.4372
0.35	1.5535
0.40	1.6910
0.45	1.8557
0.50	2.0561

10. The initial value problem is

$$y' = y - y^2, \quad y(0) = 0.5,$$

so

$$\begin{aligned} y'' &= y' - 2yy' \\ &= y - y^2 - 2y\left(y - y^2\right) \\ &= (1 - 2y)\left(y - y^2\right). \end{aligned}$$

The iteration formula is then

$$y_{n+1} = y_n + \left(y_n - y_n^2\right)h + (1 - 2y_n)\left(y_n - y_n^2\right)\frac{h^2}{2}.$$

For $h = 0.1$ we find

$$\begin{aligned} y_1 &= y_0 + \left(y_0 + y_0^2\right)(0.1) + (1 - 2y_0)\left(y_0 - y_0^2\right)\frac{(0.1)^2}{2} \\ &= 0.5 + (0.5 - 0.25)(0.1) + (1 - 1)(0.5 - 0.25)\frac{0.01}{2} \\ &= 0.5 + 0.025 \\ &= 0.525. \end{aligned}$$

The results of the remaining calculations are shown in the table. Also shown is the table obtained when $h = 0.05$.

3-Term Taylor with $h = 0.1$	
x_n	y_n
0.00	0.5000
0.10	0.5250
0.20	0.5499
0.30	0.5745
0.40	0.5988
0.50	0.6226

3-Term Taylor with $h = 0.05$	
x_n	y_n
0.00	0.5000
0.05	0.5125
0.10	0.5250
0.15	0.5374
0.20	0.5498
0.25	0.5622
0.30	0.5745
0.35	0.5866
0.40	0.5987
0.45	0.6107
0.50	0.6225

Exercises 9.4

4. In the initial value problem

$$y' = x^2 + y^2, \quad y(0) = 1$$

we identify $f(x, y) = x^2 + y^2$, $x_0 = 0$, and $y_0 = 1$. For $h = 0.1$ we compute y_1:

$$\begin{aligned}
k_1 &= hf(x_0, y_0) \\
&= 0.1(0 + 1) \\
&= 0.1 \\
k_2 &= hf\left(x_0 + \frac{1}{2}h, y_0 + \frac{1}{2}k_1\right) \\
&= 0.1f(0.05, 1.05) \\
&= 0.1(0.0025 + 1.1025) \\
&= 0.1105 \\
k_3 &= hf\left(x_0 + \frac{1}{2}h, y_0 + \frac{1}{2}k_2\right) \\
&= 0.1f(0.05, 1.05525) \\
&= 0.1(0.0025 + 1.11355) \\
&= 0.1116 \\
k_4 &= hf(x_0 + h, y_0 + k_3) \\
&= 0.1f(0.1, 1.1116) \\
&= 0.1(0.01 + 1.2357) \\
&= 0.1246 \\
y_1 &= y_0 + \frac{1}{6}(k_1 + 2k_2 + 2k_3 + k_4) \\
&= 1 + \frac{1}{6}(0.1 + 0.221 + 0.2232 + 0.1246) \\
&= 1.1115.
\end{aligned}$$

The remaining values are shown in the table.

Runge-Kutta Method with $h = 0.1$	
x_n	y_n
0.00	1.0000
0.10	1.1115
0.20	1.2530
0.30	1.4397
0.40	1.6961
0.50	2.0670

10. In the initial value problem

$$y' = y - y^2, \quad y(0) = 0.5$$

we identify $f(x, y) = y - y^2$, $x_0 = 0$, and $y_0 = 0.5$. For $h = 0.1$ we compute y_1:

$$\begin{aligned}
k_1 &= hf(x_0, y_0) \\
&= 0.1(0.5 - 0.25) \\
&= 0.025 \\
k_2 &= hf\left(x_0 + \frac{1}{2}h, y_0 + \frac{1}{2}k_1\right) \\
&= 0.1f(0.05, 0.5125) \\
&= 0.1(0.5125 - 0.2627) \\
&= 0.0250 \\
k_3 &= hf\left(x_0 + \frac{1}{2}h, y_0 + \frac{1}{2}k_2\right) \\
&= 0.1f(0.05, 0.5250) \\
&= 0.1(0.5250 - 0.2756) \\
&= 0.0249 \\
k_4 &= hf(x_0 + h, y_0 + k_3) \\
&= 0.1f(0.1, 0.5249) \\
&= 0.1(0.5249 - 0.2755) \\
&= 0.0249
\end{aligned}$$

$$
\begin{aligned}
y_1 &= y_0 + \frac{1}{6}(k_1 + 2k_2 + 2k_3 + k_4) \\
&= 0.5000 + \frac{1}{6}(0.0250 + 0.0500 + 0.0498 + 0.0249) \\
&= 0.5250.
\end{aligned}
$$

The remaining values are shown in the table.

Runge-Kutta Method with $h = 0.1$	
x_n	y_n
0.00	0.5000
0.10	0.5250
0.20	0.5498
0.30	0.5744
0.40	0.5987
0.50	0.6225

Exercises 9.5

4. Identifying $f(x, y, y') = 2y' - xy$, the Euler iteration formulas become

$$
\begin{aligned}
y_{n+1} &= y_n + (0.1)u_n \\
u_{n+1} &= u_n + (0.1)(2u_n - x_n y_n).
\end{aligned}
$$

From the initial conditions we identify $y_0 = -1$ and $u_0 = 5$. Then

$$
\begin{aligned}
y_1 &= y_0 + (0.1)u_0 = -1 + (0.1)5 = -0.5 \\
u_1 &= u_0 + (0.1)[2u_0 - x_0 y_0] = 5 + (0.1)[10] = 6 \\
y_2 &= y_1 + (0.1)u_1 = -0.5 + (0.1)6 = 0.1 \\
u_2 &= u_1 + (0.1)[2u_1 - x_1 y_1] \\
&= 6 + (0.1)[12 - 0.1(-0.5)] = 7.205 \\
y_3 &= y_2 + (0.1)u_2 = 0.1 + (0.1)(7.205) = 0.8205
\end{aligned}
$$

6. Starting with $y_0 = y + \epsilon$ in the recurrence formula

$$
y_{n+1} = k(1 - y_n)
$$

we obtain

$$
\begin{aligned}
y_1 &= k(1 - y_0) \\
&= k(1 - y - \epsilon) \\
y_2 &= k(1 - y_1) \\
&= k - ky_1 \\
&= k - k^2(1 - y - \epsilon) \\
y_3 &= k(1 - y_2) \\
&= k - ky_2 \\
&= k - k^2 + k^3(1 - y - \epsilon) \\
y_4 &= k(1 - y_3) \\
&= k - ky_3 \\
&= k - k^2 + k^3 - k^4(1 - y - \epsilon)
\end{aligned}
$$

and so on. In general, we have

$$y_n = k - k^2 + k^3 - \cdots + (-1)^{n+1}k^n(1 - y - \epsilon)$$

for $n = 1, 2, 3, \ldots$. Thus, the initial error ϵ has propagated to $k_n\epsilon$ by the nth iterate. For $|k| > 1$ this is an increasing error. For $|k| < 1$ the error decreases. Therefore, the formula is unstable for $|k| > 1$ and stable for $|k| < 1$.

10 Partial Differential Equations

Exercises 10.1

4. To show that $f_1(x) = \cos x$ and $f_2(x) = \sin^2 x$ are orthogonal on $[0, \pi]$ we compute

$$\int_0^{\pi} \cos x \sin^2 x \, dx = \left. \frac{\sin^3 x}{3} \right|_0^{\pi} = \frac{1}{3}\left[\sin^3 \pi - \sin^3 0\right] = 0.$$

Therefore, $f_1(x) = \cos x$ and $f_2(x) = \sin^2 x$ are orthogonal on $[0, \pi]$.

12. To show that the set of functions

$$\left\{1, \cos \frac{n\pi}{p}x, \sin \frac{m\pi}{p}x\right\}, \qquad n,\, m = 1,\, 2,\, 3,\, \ldots$$

is orthogonal for $-p \le x < p$ we first show that 1 and $\cos \frac{n\pi}{p}x$ are orthogonal for $n = 1, 2, 3, \ldots$.

$$\int_{-p}^{p} 1 \cdot \cos \frac{n\pi}{p}x \, dx = \left. \frac{p}{n\pi} \sin \frac{n\pi}{p}x \right|_{-p}^{p} = \frac{p}{n\pi}\left[\sin n\pi - \sin(-n\pi)\right] = 0.$$

It can be shown in a similar fashion that 1 and $\sin \frac{m\pi}{p}x$ are orthogonal. Consider now $\cos \frac{n\pi}{p}x$ and $\cos \frac{s\pi}{p}x$ for $n \neq s$.

$$\begin{aligned}
\int_{-p}^{p}\left(\cos \frac{n\pi}{p}x\right)\left(\cos \frac{s\pi}{p}x\right) dx &= \frac{1}{2}\int_{-p}^{p}\left[\cos(n+s)\frac{\pi}{p}x + \cos(n-s)\frac{\pi}{p}x\right] dx \\
&= \frac{1}{2}\left[\frac{p\sin(n+s)\pi x/p}{(n+s)\pi} + \frac{p\sin(n-s)\pi x/p}{(n-s)\pi}\right]_{-p}^{p} \\
&= 0.
\end{aligned}$$

Thus, $\cos \frac{n\pi}{p}x$ and $\cos \frac{s\pi}{p}x$ are orthogonal for $n \neq s$. It can be shown that $\sin \frac{m\pi}{p}x$ and $\sin \frac{t\pi}{p}x$ are orthogonal for $m \neq t$ by using the identity

$$\sin \frac{m\pi}{p}x \sin \frac{t\pi}{p}x = \frac{1}{2}\left[\cos(m-t)\frac{\pi}{p}x - \cos(m+t)\frac{\pi}{p}x\right].$$

Finally, using the identity

$$\sin \frac{m\pi}{p}x \cos \frac{n\pi}{p}x = \frac{1}{2}\left[\sin(m-n)\frac{\pi}{p}x + \sin(m+n)\frac{\pi}{p}x\right]$$

it can be shown that $\sin \frac{m\pi}{p}x$ and $\cos \frac{n\pi}{p}x$ are orthogonal. Therefore, the set

$$\left\{1, \cos \frac{n\pi}{p}x, \sin \frac{m\pi}{p}x\right\}, \quad n, m = 1, 2, \ldots$$

is orthogonal for $-p \le x \le p$.

Exercises 10.2

4. For the function

$$f(x) = \begin{cases} 0, & -1 < x < 0 \\ x, & 0 \le x < 1 \end{cases}$$

we compute

$$a_0 = \frac{1}{1}\int_{-1}^{1} f(x)\,dx = \int_0^1 x\,dx = \frac{1}{2}x^2\Big|_0^1 = \frac{1}{2}.$$

To find a_n and b_n we integrate by parts:

$$\begin{aligned} a_n &= \int_0^1 x\cos n\pi x\,dx = \frac{1}{n\pi}x\sin n\pi x\Big|_0^1 - \frac{1}{n\pi}\int_0^1 \sin n\pi x\,dx \\ &= 0 - \frac{1}{n^2\pi^2}\left[-\cos n\pi x\right]_0^1 = \frac{1}{n^2\pi^2}[\cos n\pi - 1] \\ &= \frac{(-1)^n - 1}{n^2\pi^2}, \\ b_n &= \int_0^1 x\sin n\pi x\,dx \\ &= -\frac{1}{n\pi}x\cos n\pi x\Big|_0^1 + \frac{1}{n\pi}\int_0^1 \cos n\pi x\,dx \\ &= -\frac{\cos n\pi}{n} - 0 + \frac{1}{n^2\pi^2}\sin n\pi x\Big|_0^1 \\ &= -\frac{(-1)^n}{n\pi} = \frac{(-1)^{n+1}}{n\pi}. \end{aligned}$$

Therefore,

$$f(x) = \frac{1}{4} + \sum_{n=1}^{\infty}\left[\frac{(-1)^n - 1}{n^2\pi^2}\cos n\pi x + \frac{(-1)^{n+1}}{n\pi}\sin n\pi x\right].$$

10. For the function

$$f(x) = \begin{cases} 0, & -\frac{\pi}{2} < x < 0 \\ \cos x, & 0 \le x < \frac{\pi}{2} \end{cases}$$

we compute

$$a_0 = \frac{1}{\pi/2}\int_0^{\pi/2} \cos x\,dx = \frac{2}{\pi}\sin x\Big|_0^{\pi/2} = \frac{2}{\pi},$$

$$a_n = \frac{1}{\pi/2}\int_0^{\pi/2} \cos x \cos\frac{n\pi}{\pi/2}x\,dx = \frac{2}{\pi}\int_0^{\pi/2}\cos x\cos 2nx\,dx$$

$$= \frac{2}{\pi}\int_0^{\pi/2}\frac{1}{2}[\cos(2n-1)x + \cos(2n+1)x]\,dx$$

$$= \frac{1}{\pi}\left[\frac{1}{2n-1}\sin(2n-1)x + \frac{1}{2n+1}\sin(2n+1)x\right]_0^{\pi/2}$$

$$= \frac{1}{\pi}\left[\frac{1}{2n-1}\sin(2n-1)\frac{\pi}{2} + \frac{1}{2n+1}\sin(2n+1)\frac{\pi}{2}\right]$$

$$= \frac{1}{\pi}\left[\frac{(-1)^{n+1}}{2n-1} + \frac{(-1)^n}{2n+1}\right] = \frac{(-1)^{n+1}}{\pi}\left[\frac{2n+1-(2n-1)}{4n^2-1}\right]$$

$$= \frac{2(-1)^{n+1}}{\pi(4n^2-1)},$$

and

$$b_n = \frac{2}{\pi}\int_0^{\pi/2}\cos x\sin 2nx\,dx$$

$$= \frac{2}{\pi}\int_0^{\pi/2}\frac{1}{2}[\sin(2n-1)x + \sin(2n+1)x]\,dx$$

$$= \frac{1}{\pi}\left[-\frac{1}{2n-1}\cos(2n-1)x - \frac{1}{2n+1}\cos(2n+1)x\right]_0^{\pi/2}$$

$$= \frac{1}{\pi}\left[\frac{1}{2n-1} + \frac{1}{2n+1}\right] = \frac{4n}{\pi(4n^2-1)}.$$

Therefore,

$$f(x) = \frac{1}{\pi} + \sum_{n=1}^{\infty}\left[\frac{2(-1)^{n+1}}{\pi(4n^2-1)}\cos 2nx + \frac{4n}{\pi(4n^2-1)}\sin 2nx\right].$$

20. From Problem 9 we have

$$\sin x = \frac{1}{\pi} - \frac{2}{\pi}\sum_{n=1}^{\infty}\left[\frac{1}{4n^2-1}\cos 2nx\right] + \frac{1}{2}\sin x.$$

Solving for $\sin x$ and multiplying by $\pi/4$ this becomes

$$\frac{\pi}{4}\sin x = \frac{1}{2} - \sum_{n=1}^{\infty}\left[\frac{1}{(2n-1)(2n+1)}\cos 2nx\right].$$

For $x = \frac{\pi}{2}$ we have

$$\frac{\pi}{4} = \frac{1}{2} + \frac{1}{1\cdot 3} - \frac{1}{3\cdot 5} + \frac{1}{5\cdot 7} - \frac{1}{7\cdot 9} + \cdots.$$

Exercises 10.3

8. For $x > 0$,

$$f(-x) = -x + 5 = f(x),$$

so $f(x)$ is an even function.

20. Since

$$f(x) = \begin{cases} x+1, & -1 < x < 0 \\ x-1, & 0 \le x < 1 \end{cases}$$

is an odd function, we expand in a sine series:

$$\begin{aligned} b_n &= 2\int_0^1 f(x)\sin n\pi x\,dx = 2\int_0^1 (x-1)\sin n\pi x\,dx \\ &= 2\left[\int_0^1 x\sin n\pi x\,dx - \int_0^1 \sin n\pi x\,dx\right] \\ &= 2\left[\frac{1}{n^2\pi^2}\sin n\pi x - \frac{x}{n\pi}\cos n\pi x + \frac{1}{n\pi}\cos n\pi x\right]_0^1 \\ &= 2\left[-\frac{1}{n\pi}(-1)^n + \frac{1}{n\pi}(-1)^n - \frac{1}{n\pi}\right] = -\frac{2}{n\pi}. \end{aligned}$$

Thus

$$f(x) = -\sum_{n=1}^{\infty} \frac{2}{n\pi}\sin n\pi x.$$

24. Since

$$f(x) = \cos x, \quad -\frac{\pi}{2} < x < \frac{\pi}{2}$$

is an even function, we expand in a cosine series: (See the solution of Problem 10 in Exercise 10.2 for the computation of the integrals.)

$$a_0 = \frac{2}{\pi/2}\int_0^{\pi/2} \cos x\,dx = \frac{4}{\pi}$$

$$a_n = \frac{2}{\pi/2}\int_0^{\pi/2} \cos x\cos\frac{n\pi}{\pi/2}x\,dx = \frac{4(-1)^{n+1}}{\pi\,(4n^2-1)}$$

Thus

$$f(x) = \frac{2}{\pi} + \sum_{n=1}^{\infty} \frac{4(-1)^{n+1}}{\pi\,(4n^2-1)}\cos 2nx.$$

26. We have

$$a_0 = \frac{2}{1}\int_0^1 f(x)\,dx = 2\int_{1/2}^1 dx = 1$$

and

$$a_n = 2\int_{1/2}^1 \cos n\pi x\,dx = \frac{2}{n\pi}\sin n\pi x\bigg|_{1/2}^1$$

$$= \frac{2}{n\pi}\left(\sin n\pi - \sin\frac{n\pi}{2}\right) = -\frac{2}{n\pi}\left[\frac{(-1)^n - 1}{2}\right]^{(n-1)/2}.$$

The half-range cosine expansion of $f(x)$ is

$$f(x) = \frac{1}{2} - \sum_{n=1}^{\infty}\frac{2}{n\pi}\left[\frac{(-1)^n - 1}{2}\right]^{(n-1)/2}\cos n\pi x.$$

Next,

$$b_n = 2\int_{1/2}^1 \sin n\pi x\,dx = -\frac{2}{n\pi}\cos n\pi x\bigg|_{1/2}^1$$

$$= -\frac{2}{n\pi}\left[\cos n\pi - \cos\frac{n\pi}{2}\right] = \frac{2}{n\pi}\left[\cos\frac{n\pi}{2} - \cos n\pi\right]$$

$$= \frac{2}{n\pi}\left\{\left[\frac{(-1)^{n+1} - 1}{2}\right]^{n/2} - (-1)^n\right\}.$$

The half-range sine expansion of $f(x)$ is

$$f(x) = \frac{2}{n\pi}\sum_{n=1}^{\infty}\left\{\left[\frac{(-1)^{n+1} - 1}{2}\right]^{n/2} - (-1)^n\right\}\sin n\pi x.$$

36. We define f to be periodic with period π. Then, with $p = \pi/2$, we have

$$a_0 = \frac{1}{\pi/2}\int_0^\pi x\,dx = \frac{2}{\pi}\left[\frac{x^2}{2}\right]_0^\pi = \pi,$$

$$a_n = \frac{1}{\pi/2}\int_0^\pi x\cos 2nx\,dx = \frac{2}{\pi}\left[\frac{1}{4n^2}\cos 2nx + \frac{x}{2n}\sin 2nx\right]_0^\pi$$

$$= \frac{2}{\pi}\left[\frac{1}{4n^2} - \frac{1}{4n^2}\right] = 0,$$

and

$$b_n = \frac{1}{\pi/2}\int_0^\pi x\sin 2nx\,dx = \frac{2}{\pi}\left[\frac{1}{4n^2}\sin 2nx - \frac{x}{2n}\cos 2nx\right]_0^\pi$$

$$= \frac{2}{\pi}\left[-\frac{\pi}{2n}\right] = -\frac{1}{n}.$$

Hence

$$f(x) = \frac{\pi}{2} - \sum_{n=1}^{\infty} \frac{1}{n} \sin 2nx.$$

Exercises 10.4

6. Treating x as a constant we write

$$\frac{\partial^2 u}{\partial y^2} = -\sin xy.$$

Integrating, we have

$$\frac{\partial u}{\partial x} = -\int \sin xy\, dy = \frac{1}{x} \cos xy + f(x)$$

and

$$u(x,y) = \int \frac{1}{x} \cos xy\, dy + \int f(x)\, dy$$

$$= \frac{1}{x^2} \sin xy + f(x)y + g(x).$$

14. If we let $v = \partial u / \partial y$, the equation becomes

$$y \frac{\partial v}{\partial y} + v = 0.$$

Treating v as a function of the single variable y we separate variables:

$$\frac{dv}{v} = -\frac{dy}{y}$$

$$\ln|v| = -\ln|y| + c = -\ln|c_1 y|$$

$$v = \frac{1}{c_2 y}.$$

Letting $c_2 = f(x)$ we have

$$\frac{\partial u}{\partial y} = v = \frac{1}{f(x)y},$$

so

$$u = \frac{\ln|y|}{f(x)} + g(x).$$

From the given conditions, we obtain

$$x^2 = u(x,1) = \frac{\ln 1}{f(x)} + g(x) = g(x)$$

and

$$1 = u(x, e) = \frac{\ln e}{f(x)} x^2 = \frac{1}{f(x)} + x^2.$$

Thus, $f(x) = \frac{1}{1-x^2}$, and the solution of the partial differential equation subject to the given conditions is

$$u(x, y) = \left(1 - x^2\right) \ln |y| + x^2.$$

16. Letting $u(x, y) = X(x)Y(y)$ the differential equation can be written

$$X'Y + 3XY' = 0.$$

Separating variables, we find

$$\frac{X'}{X} = -3\frac{Y'}{Y}.$$

We consider three cases:

(I) $\lambda^2 > 0$. We have the ordinary differential equations

$$X' - \lambda^2 X = 0 \qquad \text{and} \qquad Y' + \frac{1}{3}\lambda^2 Y = 0.$$

These have general solutions

$$X = c_1 e^{\lambda^2 x} \qquad \text{and} \qquad Y = c_2 e^{-\lambda^2 y/3}.$$

Thus, a particular solution of the partial differential equation is

$$u(x, y) = A_1 e^{\lambda^2(x - y/3)}$$

(II) $-\lambda^2 < 0$. We have the ordinary differential equations

$$X' + \lambda^2 X = 0 \qquad \text{and} \qquad Y' - \frac{1}{3}\lambda^2 Y = 0.$$

These have general solutions

$$X = c_3 e^{-\lambda^2 x} \qquad \text{and} \qquad Y = c_4 e^{\lambda^2 y/3}.$$

Thus, another particular solution is

$$u(x, y) = A_2 e^{\lambda^2(-x + y/3)}.$$

(III) $\lambda^2 = 0$. We have

$$X' = 0 \qquad \text{and} \qquad Y' = 0.$$

Then,

$$X = c_5 \qquad \text{and} \qquad Y = c_6$$

so that

$$u(x, y) = A_3.$$

Then,

$$X = c_5 \qquad \text{and} \qquad Y = c_6$$

so that

$$u(x, y) = A_3.$$

22. Letting $u(x, y) = X(x)Y(y)$ the differential equation can be written

$$yX'Y' + XY = 0.$$

Separating variables we find

$$y\frac{Y'}{Y} = -\frac{X}{X'}.$$

We consider three cases:

(I) $\lambda^2 > 0$. The separated ordinary differential equations

$$\lambda^2 X' + X = 0 \qquad \text{and} \qquad yY' - \lambda^2 Y = 0$$

have the general solutions

$$X = c_1 e^{-x/\lambda^2} \qquad \text{and} \qquad Y = c_2 y^{\lambda^2}.$$

Hence, a particular solution of the given partial differential equation is

$$u = A_1 y^{\lambda^2} e^{-x/\lambda^2}.$$

(II) $-\lambda^2 < 0$. The separated ordinary differential equations are now

$$\lambda^2 X' - X = 0 \qquad \text{and} \qquad yY' + \lambda^2 Y = 0.$$

These have general solutions

$$X = c_3 e^{x/\lambda^2} \qquad \text{and} \qquad Y = c_4 y^{-\lambda^2}.$$

A particular solution of the partial differential is then

$$u = A_2 y^{-\lambda^2} e^{x/\lambda^2}.$$

(III) $\lambda^2 = 0$. In this case we have $X = 0$ and $Y' = 0$. Thus we see a particular solution is $u = 0$.

28. Letting $u(x, y) = X(x)Y(y)$ the differential equation can be written as

$$x^2 X''Y + XY'' = 0.$$

Separating variables, we find

$$x^2\frac{X''}{X} = -\frac{Y''}{Y}.$$

We consider three cases:

(I) $\lambda^2 > 0$. We have

$$x^2X'' - \lambda^2 X = 0 \qquad \text{and} \qquad Y'' + \lambda^2 Y = 0.$$

The second equation has the general solution

$$Y = c_1 \sin \lambda y + c_2 \cos \lambda y.$$

The first equation is a Cauchy-Euler equation. Letting $X = x^m$ we find

$$m = \frac{1 \pm \sqrt{1+4\lambda^2}}{2}.$$

Thus,

$$X = c_3 x^{(1+\sqrt{1+4\lambda^2})/2} + c_4 x^{(1-\sqrt{1+4\lambda^2})/2}$$

and a particular solution is

$$\begin{aligned} u(x,y) &= \left[c_3 x^{(1+\sqrt{1+4\lambda^2})/2} c_4 x^{(1-\sqrt{1+4\lambda^2})/2}\right] [c_1 \sin \lambda y + c_2 \cos \lambda y] \\ &= A_1 x^{(1+\sqrt{1+4\lambda^2})/2} \sin \lambda y + A_2 x^{(1-\sqrt{1+4\lambda^2})/2} \sin \lambda y \\ &\qquad + A_3 x^{(1+\sqrt{1+4\lambda^2})/2} \cos \lambda y + A_4 x^{(1-\sqrt{1+4\lambda^2})/2} \cos \lambda y. \end{aligned}$$

(II) $-\lambda^2 < 0$. We have

$$x^2X'' + \lambda^2 X = 0 \qquad \text{and} \qquad Y'' + \lambda^2 Y = 0.$$

The second equation has the general solution

$$Y = c_5 e^{\lambda y} + c_6 e^{-\lambda y}.$$

Letting $X = x^m$ in the first equation, we find

$$m = \frac{1 \pm \sqrt{1-4\lambda^2}}{2}.$$

We need to consider three cases:

(a) $1 - 4\lambda^2 > 0$. In this case $-1/2 < \lambda < 1/2$ and the values of m are real. We thus have

$$X = c_7 x^{(1+\sqrt{1-4\lambda^2})/2} + c_8 x^{(1-\sqrt{1-4\lambda^2})/2}.$$

A particular solution of the partial differential eqation is

$$u = A_5 x^{(1+\sqrt{1-4\lambda^2})/2} e^{\lambda y} + A_6 x^{(1-\sqrt{1-4\lambda^2})/2} e^{\lambda y}$$

$$+ A_7 x^{(1+\sqrt{1-4\lambda^2})/2} e^{-\lambda y} + A_8 x^{(1-\sqrt{1-4\lambda^2})/2} e^{-\lambda y}.$$

(b) $1 - 4\lambda^2 = 0$. In this case $\lambda = \pm 1/2$ and m is the double root $1/2$. Thus,

$$X = c_9 \sqrt{x} + c_{10} \sqrt{x} \ln x$$

and a particular solution is

$$u = A_9 \sqrt{x}\, e^{y/2} + A_{10} \sqrt{x}\, e^{-y/2} + A_{11} \sqrt{x}\, e^{y/2} \ln x + A_{12} \sqrt{x}\, e^{-y/2} \ln x.$$

(c) $1 - 4\lambda^2 < 0$. In this case $\lambda < -1/2$ or $\lambda > 1/2$ and the values of m are complex. We thus have

$$X = c_{11} \sqrt{x} \cos\left(\ln x^{\sqrt{4\lambda^2-1}/2}\right) + c_{12} \sqrt{x} \sin\left(\ln x^{\sqrt{4\lambda^2-1}/2}\right)$$

and so another particular solution is given by

$$u(x,y) = A_{13} \sqrt{x} \cos\left(\ln x^{\sqrt{4\lambda^2-1}/2}\right) e^{\lambda y} + A_{14} \sqrt{x} \sin\left(\ln x^{\sqrt{4\lambda^2-1}/2}\right) e^{\lambda y}$$

$$+ A_{15} \sqrt{x} \cos\left(\ln x^{\sqrt{4\lambda^2-1}/2}\right) e^{-\lambda y} + A_{16} \sqrt{x} \sin\left(\ln x^{\sqrt{4\lambda^2-1}/2}\right) e^{-\lambda y}.$$

(III) $\lambda^2 = 0$. we have

$$X'' = 0 \quad \text{and} \quad Y'' = 0.$$

Then,

$$X = c_{13} x + c_{14} \quad \text{and} \quad Y = c_{15} y + c_{16}$$

so that

$$u(x,y) = A_{17} xy + A_{18} x + A_{19} y + A_{20}.$$

34. Letting $u(x,y) = X(x)Y(y)$ the differential equation can be written in the form

$$X''Y + Y''X = 0. \tag{1}$$

The given conditions become

$$X'(0)Y(y) = 0, \qquad X'(\pi)Y(y) = 0, \qquad X(x)Y(0) = 0;$$

or

$$X'(0) = 0, \qquad X'(\pi) = 0, \qquad Y(0) = 0.$$

Separating variables in (1) we obtain

$$\frac{X''}{X} = -\frac{Y''}{Y}.$$

We consider three cases:

(I) $\lambda^2 > 0$. We have

$$X'' - \lambda^2 X = 0 \qquad \text{and} \qquad Y'' + \lambda^2 Y = 0,$$

with corresponding solutions

$$X = c_1 \cosh \lambda x + c_2 \sinh \lambda x \qquad \text{and} \qquad Y = c_3 \cos \lambda y + c_4 \sin \lambda y.$$

Since

$$X' = c_1 \lambda \sinh \lambda x + c_2 \lambda \cosh \lambda x$$

the boundary conditions imply

$$c_1 \lambda \sinh 0 + c_2 \lambda \cosh 0 = 0 \qquad \text{and} \qquad c_1 \lambda \sinh \lambda\pi + c_2 \lambda \cosh \lambda\pi = 0$$

or

$$c_2 \lambda = 0 \qquad \text{and} \qquad (c_1 \sinh \lambda\pi + c_2 \cosh \lambda\pi)\lambda = 0.$$

Since $\lambda > 0$, $c_2 = 0$ and $c_1 \sinh \lambda\pi = 0$, so $c_1 = 0$. Thus $X(x) = 0$ and $u(x, y) = 0$.

(II) $-\lambda^2 < 0$. We have

$$X'' + \lambda^2 X = 0 \qquad \text{and} \qquad Y'' - \lambda^2 Y = 0$$

with corresponding solutions

$$X = c_1 \cos \lambda x + c_2 \sin \lambda x \qquad \text{and} \qquad Y = c_3 \cosh \lambda y + c_4 \sinh \lambda y.$$

Since

$$X' = -c_1 \lambda \sin \lambda x + c_2 \lambda \cos \lambda x$$

the boundary conditions imply

$$-c_1 \lambda \sin 0 + c_2 \lambda \cos 0 = 0 \qquad \text{and} \qquad -c_1 \lambda \sin \lambda\pi + c_2 \lambda \cos \lambda\pi = 0$$

or

$$c_2 \lambda = 0 \qquad \text{and} \qquad -c_1 \lambda \sin \lambda\pi + c_2 \lambda \cos \lambda\pi = 0.$$

Since $\lambda > 0$, $c_2 = 0$ and $c_1 \sin \lambda\pi = 0$. To avoid the trivial solution we take $c_1 \neq 0$ and let $\sin \lambda\pi = 0$. This implies that λ is an integer. Thus,

$$X(x) = C_n \cos nx,$$

where we may assume that n is a positive integer [since $\cos(-t) = \cos t$]. From $Y = c_3 \cosh ny + c_4 \sinh ny$ and $Y(0) = 0$ we conclude that $c_3 = 0$. Thus $Y(y) = c_4 \sinh ny$ and

$$u_n(x, y) = A_n \cos nx \sinh ny,$$

for n a positive integer.

(III) $\lambda = 0$. We have

$$X'' = 0 \quad \text{and} \quad Y'' = 0$$

with corresponding solutions

$$X = ax + b \quad \text{and} \quad Y = cy + d.$$

From $X'(x) = a$ and $X'(0) = 0$ we obtain $a = 0$ and $X(x) = b$ [which also satisfies $X'(\pi) = 0$]. From $Y = cy + d$ and $Y(0) = 0$ we obtain $d = 0$ and $Y(y) = cy$. Thus, $u(x, y) = A_0 y$.

Using the superposition principle we have the following solution to the boundary-value problem:

$$u(x, y) = A_0 + \sum_{n=1}^{\infty} A_n \cos nx \sinh ny.$$

Exercises 10.5

6. Letting $u(x, y) = X(x)T(t)$ and separating variables we obtain

$$\frac{X''}{X} = \frac{T' + hT}{kT} = -\lambda^2.$$

The separated equations are

$$X'' + \lambda^2 X = 0 \quad \text{and} \quad T' + \left(h + k\lambda^2\right) T = 0$$

whose solutions are

$$X = c_1 \cos \lambda x + c_2 \sin \lambda x \quad \text{and} \quad T = c_3 e^{-\left(h + k\lambda^2\right)t}.$$

The boundary conditions $u(0, t) = 0$ and $u(L, t) = 0$ yield $X(0) = 0$ and $X(L) = 0$. The condition $X(0) = 0$ then implies $c_1 = 0$. Hence we have $X(x) = c_2 \sin \lambda x$. Now the second condition $X(L) = 0$ is satisfied when $c_2 \sin \lambda L = 0$. For a nontrivial solution we must have $\sin \lambda L = 0$ or $\lambda L = n\pi$, $n = 1, 2, 3, \ldots$. The eigenvalues and eigenfunctions are

$$\lambda = \frac{n\pi}{L} \quad \text{and} \quad X(x) = c_2 \sin \frac{n\pi}{L} x, \quad n = 1, 2, 3, \ldots.$$

Product solutions that satisfy the partial differential equation and boundary conditions are

$$u_n(x,t) = XT = A_n e^{-(h+kn^2\pi^2/L^2)t} \sin\frac{n\pi}{L}x. \tag{1}$$

In order to obtain a solution that satisfies the initial condition we use the superposition principle:

$$u(x,t) = e^{-ht}\sum_{n=1}^{\infty} A_n e^{-(kn^2\pi^2/L^2)t} \sin\frac{n\pi}{L}x.$$

At $t = 0$, $u(x,0) = f(x)$ and so

$$f(x) = \sum_{n=1}^{\infty} A_n \sin\frac{n\pi}{L}x$$

implies

$$A_n = \frac{2}{L}\int_0^L f(x)\sin\frac{n\pi}{L}x\,dx. \tag{2}$$

A solution of the original problem consists of the series (1) with the coefficients A_n defined by (2).

10. The separated equations

$$X'' + \lambda^2 X = 0 \qquad \text{and} \qquad T'' + a^2\lambda^2 T = 0$$

have the general solutions

$$X(x) = c_1\cos\lambda x + c_2\sin\lambda x \qquad \text{and} \qquad T(t) = c_3\cos a\lambda t + c_4\sin a\lambda t.$$

The homogeneous conditions $u(0,t) = 0$, $u(\pi,t) = 0$, and $u_t(x,0) = 0$ are equivalent to $X(0) = 0$, $X(\pi) = 0$, and $T'(0) = 0$, respectively. Applying the first and third conditions to the appropriate solutions gives, in turn, $c_1 = 0$ and $c_4 = 0$. Hence

$$X(x) = c_2\sin\lambda x \qquad \text{and} \qquad T(t) = c_3\cos a\lambda t.$$

Finally, $X(\pi) = 0$ yields $\sin\lambda\pi = 0$ which implies that the eigenvalues are $\lambda = n$; $n = 1, 2, 3, \ldots$. Forming the product XT and using the superposition principle gives

$$u(x,t)\sum_{n=1}^{\infty} A_n\cos ant\sin nx.$$

At $t = 0$ we must have

$$\frac{1}{6}x\left(\pi^2 - x^2\right) = \sum_{n=1}^{\infty} A_n\sin nx$$

and so

$$A_n = \frac{2}{\pi}\cdot\frac{1}{6}\int_0^\pi\left(\pi^2 x - x^3\right)\sin nx\,dx.$$

Integrating by parts yields

$$A_n = \frac{2(-1)^{n+1}}{n^3}.$$

Therefore a solution to the boundary-value problem is

$$U(x,t) = 2\sum_{n=1}^{\infty} \frac{(-1)^{n+1}}{n^3} \cos ant \sin nx.$$

14. Letting $u(x,t) = X(x)T(t)$ and separating variables we obtain

$$\frac{X''}{X} = \frac{T'' + 2\beta T'}{T}$$

or

$$X'' + \lambda^2 X = 0 \qquad \text{and} \qquad T'' + 2\beta T' + \lambda T = 0,$$

with corresponding solutions

$$X = c_1 \cos \lambda x + c_2 \sin \lambda x \qquad \text{and} \qquad T = e^{-\beta t}\left(c_1 \cos \sqrt{n^2 - \beta^2}\, t + c_2 \sin \sqrt{n^2 - \beta^2}\, t\right).$$

The boundary conditions $u(0,t) = 0$ and $u(\pi,t) = 0$ yield $X(0) = 0$ and $X(\pi) = 0$. The condition $x(0) = 0$ implies $c_1 = 0$, so $X(x) = c_2 \sin \lambda x$. The second condition $X(\pi) = 0$ yields $\sin \lambda\pi = 0$, so $\lambda = n$ for $n = 1, 2, 3, \ldots$. The eigenvalues and eigenfunctions are then

$$\lambda = n \qquad \text{and} \qquad X(x) = c_2 \sin nx, \quad n = 1, 2, 3, \ldots.$$

Product solutions that satisfy the partial differential equation and boundary conditions are

$$U_n(x,t) = XT = A_n e^{-\beta t}\left(c_1 \cos \sqrt{n^2 - \beta^2}\, t + c_2 \sin \sqrt{n^2 - \beta^2}\, t\right) \sin nx.$$

In order to obtain a solution that satisfies the initial condition we use the superposition principle:

$$u(x,t) = e^{-\beta t} \sum_{n=1}^{\infty} \left(A_n \cos \sqrt{n^2 - \beta^2}\, t + B_n \sin \sqrt{n^2 - \beta^2}\, t\right) \sin nx.$$

At $t = 0$, $u(x,0) = f(x)$ and so

$$f(x) = \sum_{n=1}^{\infty} A_n \sin nx,$$

which implies

$$A_n = \frac{2}{\pi} \int_0^{\pi} f(x) \sin nx.$$

To find the coefficients B_n we compute

$$\frac{\partial u}{\partial t} = e^{-\beta t} \sum_{n=1}^{\infty} \left(-A_n \sqrt{n^2 - \beta^2} \sin \sqrt{n^2 - \beta^2}\, t + B_n \sqrt{n^2 - \beta^2} \cos \sqrt{n^2 - \beta^2}\, t\right) \sin nx$$

$$- \beta e^{-\beta t} \sum_{n=1}^{\infty} \left(A_n \cos \sqrt{n^2 - \beta^2}\, t + B_n \sin \sqrt{n^2 - \beta^2}\, t\right) \sin nx.$$

Then

$$0 = \left.\frac{\partial u}{\partial t}\right|_{t=0} = \sum_{n=1}^{\infty} B_n\sqrt{n^2-\beta^2}\,\sin nx - \beta\sum_{n=1}^{\infty} A_n \sin nx$$

$$0 = \sum_{n=1}^{\infty}\left(B_n\sqrt{n^2-\beta^2} - \beta A_n\right)\sin nx$$

$$B_n\sqrt{n^2-\beta^2} - \beta A_n = 0$$

$$B_n = \frac{\beta}{\sqrt{n^2-\beta^2}}A_n.$$

Thus, the displacement is given by

$$u(x,t) = e^{-\beta t}\sum_{n=1}^{\infty} A_n\left(\cos\sqrt{n^2-\beta^2}\,t + \frac{\beta}{\sqrt{n^2-\beta^2}}\sin\sqrt{n^2-\beta^2}\,t\right)\sin nx,$$

where

$$A_n = \frac{2}{\pi}\int_0^{\pi} f(x)\sin nx.$$

18. Letting $u(x,y) = X(x)Y(y)$ and separating variables, we obtain

$$X'' + \lambda^2 X = 0 \qquad \text{and} \qquad Y'' - \lambda^2 Y = 0,$$

whose solutions are

$$X = c_1\cos\lambda x + c_2\sin\lambda x \qquad \text{and} \qquad Y = c_3\cosh\lambda y + c_4\sinh\lambda y.$$

The boundary conditions $\frac{\partial u}{\partial x}\big|_{x=0} = 0$ and $\frac{\partial u}{\partial x}\big|_{x=a} = 0$ are equivalent to $X'(0) = 0$ and $X'(a) = 0$, respectively. Applying the first of these conditions immediately gives $c_2 = 0$. Hence

$$X(x) = c_1\cos\lambda x.$$

Now $X'(a) = 0$ is the same as

$$-c_1\lambda\sin\lambda a = 0$$

which in turn implies $\lambda = 0$ or $\lambda a = n\pi$ for $n = 1, 2, 3, \ldots$.

When $\lambda = 0$ the separated equations $X'' = 0$ and $Y'' = 0$ yield $X(x) = c_1x + c_2$ and $Y(y) = c_3y + c_4$. The boundary conditions $X'(0) = 0$ and $X'(a) = 0$ are both satisfied when $c_1 = 0$. Therefore $X(x) = c_2$. Also, $Y(b) = 0$ implies $c_3b + c_4 = 0$ or $c_4 = -c_3b$. Thus $Y(y) = c_3(y - b)$. When the condition $Y(b) = 0$ is applied to $Y(y) = c_3\cosh\lambda y + c_4\sinh\lambda y$ we find

$$c_3\cosh\lambda b + c_4\sinh\lambda b = 0$$

or

$$c_4 = -c_3 \frac{\cosh \lambda b}{\sinh \lambda b}.$$

Therefore

$$Y(y) = c_3 \left[\cosh \lambda y - \frac{\cosh \lambda b}{\sinh \lambda b} \sinh \lambda y \right]$$

$$= c_3 \left[\frac{\sinh \lambda b \cosh \lambda y - \cosh \lambda b \sinh \lambda y}{\sinh \lambda b} \right]$$

$$= c_3 \frac{\sinh \lambda (b - y)}{\sinh \lambda b}.$$

For $\lambda = 0$ the product solutions are

$$A_0(y - b),$$

and for $\lambda = n\pi/a$, $n = 1, 2, 3, \ldots$ the product solutions are

$$A_n \frac{\sinh \frac{n\pi}{a}(b - y)}{\sinh \frac{n\pi}{a} b} \cos \frac{n\pi}{a} x.$$

The superposition principle then yields

$$u(x, y) = A_0(y - b) + \sum_{n=1}^{\infty} A_n \frac{\sinh \frac{n\pi}{a}(b - y)}{\sinh \frac{n\pi}{a} b} \cos \frac{n\pi}{a} x.$$

At $y = 0$ we have

$$x = -A_0 b + \sum_{n=1}^{\infty} A_n \cos \frac{n\pi}{a} x,$$

which is a half-range expansion of x in a cosine series. Using (1) in Section 10.3 of the text, we identify

$$\frac{a_0}{2} = -A_0 b \qquad \text{and} \qquad a_n = A_n.$$

We compute

$$A_0 = -\frac{a_0}{2b} = -\frac{1}{2b} \frac{2}{a} \int_0^a x\, dx = -\frac{1}{ab} \left(\frac{a^2}{2} \right) = -\frac{1}{2b}$$

and

$$a_n = \frac{2}{a} \int_0^a x \cos \frac{n\pi}{a} x\, dx$$

$$= \frac{2}{a} \left[\frac{a^2}{n^2\pi^2} \cos \frac{n\pi}{a} x + \frac{ax}{n\pi} \sin \frac{n\pi}{a} x \right]_0^a$$

$$= \frac{2}{a} \left[\frac{a^2}{n^2\pi^2} \cos n\pi - \frac{a^2}{n^2\pi^2} \right]$$

$$= \frac{2a}{n^2\pi^2} \left[(-1)^n - 1 \right].$$

Therefore, a solution of the boundary value problem is

$$u(x,y) = \frac{a}{2b}(b-y) + \frac{2a}{\pi^2}\sum_{n=1}^{\infty} \frac{(-1)^n - 1}{n^2} \frac{\sinh \frac{n\pi}{a}(b-y)}{\sinh \frac{n\pi}{a}b} \cos \frac{n\pi}{a}x.$$

22. The problem is

$$\frac{\partial^2 u}{\partial x^2} + \frac{\partial^2 u}{\partial y^2} = 0, \quad 0 < x < \pi, \quad y > 0$$

$$\left.\frac{\partial u}{\partial x}\right|_{x=0} = 0, \quad \left.\frac{\partial u}{\partial x}\right|_{x=\pi} = 0, \quad y > 0$$

$$u(x,0) = f(x), \quad 0 < x < \pi$$

$$|u(x,y)| \le M \quad \text{as} \quad y \to \infty.$$

The usual separation of variables leads to

$$X(x) = c_1 \cos \lambda x + c_2 \sin \lambda x \qquad \text{and} \qquad Y(y) = c_3 e^{-\lambda y} + c_4 e^{\lambda y}.$$

We use the latter solution of the separated equation $Y'' - \lambda^2 Y = 0$ instead of $Y(y) = c_3 \cosh \lambda y + c_4 \sinh \lambda y$ since both $\cosh \lambda y$ and $\sinh \lambda y$ are unbounded as $y \to \infty$. Now $X'(0) = 0$ implies $c_2 = 0$, whereas $X'(\pi) = 0$ implies $\lambda \sin \lambda\pi = 0$. Thus $\lambda = 0$ or $\lambda = n$; $n = 1, 2, 3, \ldots$. For $\lambda = 0$ the solutions of $X'' = 0$ and $Y'' = 0$ are in turn $X(x) = c_1 x + c_2$ and $Y(y) = c_3 y + c_4$. The conditions $X'(0) = 0$ and $X'(\pi) = 0$ imply $X(x) = c_2$. In order that $Y(y)$ be bounded as $y \to \infty$ it is necessary to define $c_3 = 0$. Hence $Y(y) = c_4$.

For $\lambda = n > 0$ it is necessary to define $c_4 = 0$. In other words, $Y(y) = c_3 e^{-ny}$ is bounded as $y \to \infty$.

For $\lambda = 0$ a product solution is A_0 where A_0 is defined as $c_2 c_4$. For $\lambda = n$, product solutions are $A_n e^{-ny} \cos nx$ wheare A_n is taken as $c_1 c_3$.

By the superposition principle, a formal solution satisfying the partial differential equation and the conditions of insulation is:

$$u(x,y) = A_0 + \sum_{n=1}^{\infty} A_n e^{-ny} \cos nx.$$

To satisfy the remaining condition at $y = 0$ we must have

$$f(x) = A_0 + \sum_{n=1}^{\infty} A_n \cos nx,$$

and consequently

$$A_0 = \frac{a_0}{2} = \frac{1}{\pi}\int_0^{\pi} f(x)\, dx$$

and

$$A_n = a_n = \frac{2}{\pi}\int_0^\pi f(x)\cos nx\,dx.$$

The steady-state temperature is given by

$$u(x,y) = A_0 + \sum_{n=1}^{\infty} A_n e^{-ny}\cos nx$$

where

$$A_0 = \frac{1}{\pi}\int_0^\pi f(x)\,dx \qquad \text{and} \qquad A_n = \frac{2}{\pi}\int_0^\pi f(x)\cos nx\,dx.$$